CliffsNotes® STAAR® EOC Biology Quick Review

CliffsNotes® STAAR® EOC Biology Quick Review

By Courtney Mayer, M.Ed.

About the Author
Courtney Mayer has taught science for 21 years and is a College Board Consultant in the Southwestern Region; she conducts workshops throughout the United States. Courtney has co-authored numerous textbooks and test-preparation guides. She was honored by the College Board in 2011 with the AP Award. Today, Courtney is the Director of GT Programs and Advanced Academics for Northside ISD in San Antonio, Texas.

Editorial
Executive Editor: Greg Tubach
Senior Editor: Christina Stambaugh
Copy Editor: Elizabeth Kuball
Technical Editors: Kellie Ploeger Cox and Scott Ryan
Proofreader: Lynn Northrup

CliffsNotes® STAAR® EOC Biology Quick Review

ISBN: 9781957671246

Table of Contents

INTRODUCTION

CliffsNotes® STAAR® EOC Biology Quick Review is a reference tool that will help you review the important elements of biology necessary to master the STAAR End of Course (EOC) Exam.

What Is STAAR End of Course?

STAAR stands for "State of Texas Assessments of Academic Readiness," and *EOC* stands for "end of course." Texas high school students must meet a Satisfactory Academic Performance (Level II) on each EOC assessment they take in order to graduate.

Students will be tested in five areas:

- Cell Structure and Function (chapter 1)
- Mechanisms of Genetics (chapter 2)
- Biological Evolution and Classification (chapter 3)
- Biological Processes and Systems (chapter 4)
- Interdependence within Environmental Systems (chapter 5)

Important facts about the STAAR EOC Biology Exam

- The exam consists of 54 multiple-choice questions.
- Students will have 4 hours to complete the STAAR EOC Biology Exam. Students are allowed to take breaks to get a drink of water, have a snack, or use the restroom, but the test clock will not pause for these breaks. Practice timing yourself while taking the two full-length practice tests in chapters 6 and 7 to best prepare yourself for the timed setting of the actual exam.

- Students will have access to a handheld, four-function, scientific or graphing calculator. There will be at least one calculator for every five students on the day of the test. If a calculator is used by multiple students, its memory must be cleared after each use.
- The test may be administered as a paper-and-pencil exam or on a computer. The test administration method varies by school and by school district.
- No formula chart is allowed during the administration of the test; any necessary information will be embedded in the test questions.

For additional information on the STAAR EOC Biology Exam, visit http://tea.texas.gov/student.assessment/staar/science.

How to Use This Book

There are a number of ways you can use this book to prepare for the STAAR EOC Biology Exam. You decide what works best for your needs. Here are a few suggested approaches:

- **Approach 1:**
    - Take Practice Test 1 (chapter 6) as a diagnostic test to see what areas you need to focus on most. Check your answers and identify the topics of the questions you missed.
    - Review these topics in chapters 1–5.
    - Answer the Chapter Check-Out questions at the end of each chapter to make sure you understand the content.
    - Take Practice Test 2 (chapter 7) for final practice.
- **Approach 2:**
    - Read chapters 1–5.
    - Answer the Chapter Check-Out questions at the end of each chapter to make sure you understand the content.
    - Take Practice Test 1 (chapter 6) for practice.
    - Take Practice Test 2 (chapter 7) for additional practice.

- **Approach 3:**
  - Answer the Chapter Check-Out questions at the end of each chapter. Review the related chapter content for any questions you missed.
  - Take Practice Test 1 (chapter 6) for practice.
  - Take Practice Test 2 (chapter 7) for additional practice.

Chapter 1

CELL STRUCTURE AND FUNCTION

Chapter Check-In

- ❑ Prokaryotic and eukaryotic cells
- ❑ Cellular processes
- ❑ Viruses
- ❑ Cell cycle and mitosis
- ❑ Biomolecules

Cells were unknown to us for many centuries. You and every other living thing are made of cells. Cells are the smallest unit of life and have different structures and functions depending on where in the body they are found or what organism they are found in. In this chapter, we will look at the history of the cell and cell theory. We will also study cellular structure, function, and transport. We will learn how cells divide to make new cells and which biomolecules are needed for each cell part to function. We will also look at viruses, which, though not technically alive, can affect cells and our bodies.

Prokaryotic and Eukaryotic Cells

History of the cell and cell theory

In 1665, Robert Hooke looked under a compound microscope at a piece of cork, a type of plant material found in the bark of an oak tree. He noticed that there seemed to be tiny compartments that looked to him like the rooms that are found in a monastery that monks lived in. In a monastery, these small rooms were called cells and, thus, the name of the cell was born. About the same time, a Dutch biologist named Anton van Leeuwenhoek had developed a more powerful microscope and was looking at organisms he found swimming in pond water. We now know these to be protozoans, although, at the time, he called them "animalcules."

In 1838, German scientist Matthias Schleiden was studying plant cells under a microscope and noted that all plants were made of cells. About the same time, another German scientist, Theodor Schwann, concluded that all animals are made of cells. In 1855, a third German scientist, Rudolf Virchow, discovered that all cells come from preexisting cells.

These discoveries led to the **cell theory,** which states:

- All living things are composed of cells.
- Cells are the basic units of structure and function in living things.
- New cells are produced from existing cells.

Cells can be put into two categories: those with a **nucleus** and membrane-bound **organelles** and those without. The nucleus contains the genetic material for the cell, and the organelles are the structures within the cell that perform specific functions. (We will get into the organelles and nucleus in the "Cell structure and function" section later in this chapter.)

Prokaryotic cells

Prokaryotic cells are simple cells that lack membrane-bound organelles. The genetic material for prokaryotic cells is just floating around in the cytoplasm and not bound in a nucleus. All bacteria are prokaryotic. Below is a picture of a prokaryotic cell (Figure 1-1). Can you see how the DNA is located in a "nucleoid region" and not in its own organelle? Also, do you notice the lack of membrane-bound organelles? The simple nature of the prokaryotic cell suggests to scientists that the first cells to evolve were prokaryotic.

Figure 1-1 Prokaryotic cell.

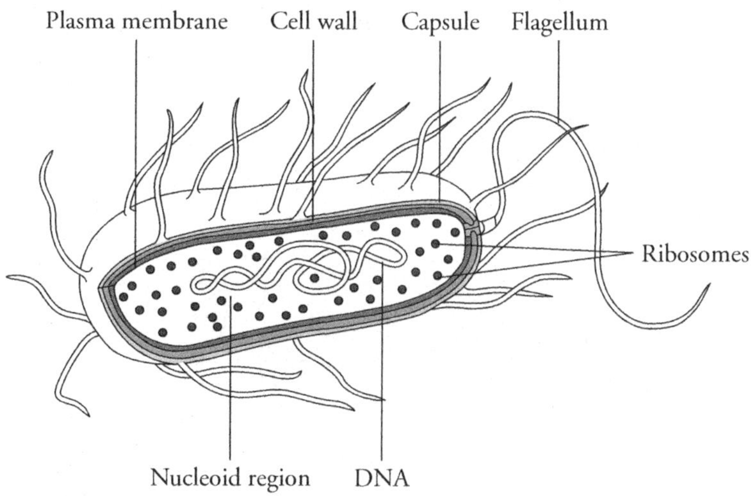

Eukaryotic cells

Eukaryotic cells, on the other hand, are more complex than prokaryotic cells. A eukaryotic cell has a nucleus containing the cell's genetic material, as well as membrane-bound organelles. All animal and plant cells are eukaryotic, as are the cells of all protists and fungi. Let's look at the cell parts found in a eukaryotic cell. Figure 1-2 is an example of a typical animal cell.

Figure 1-2 Animal cell.

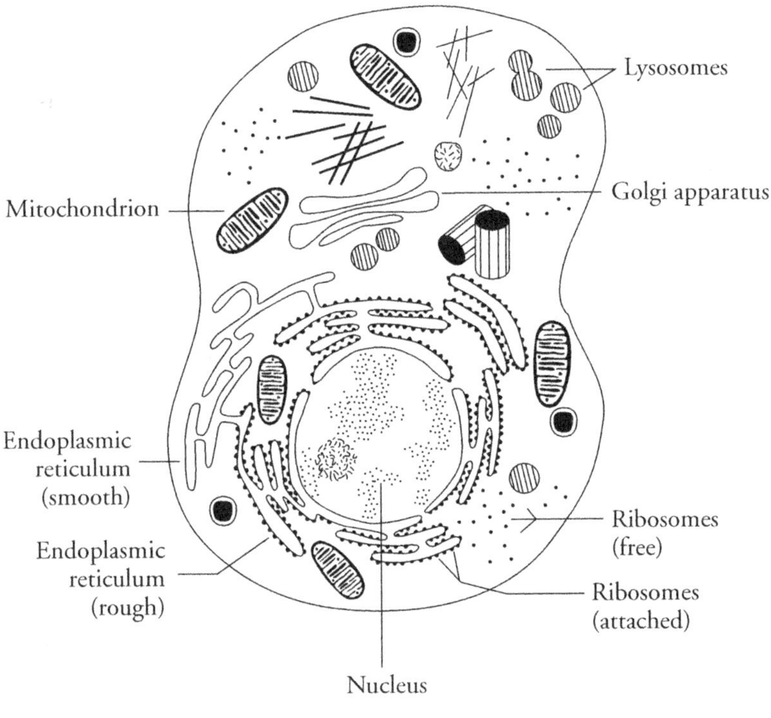

We will go through each organelle and the function/role it plays in the cell. To do this, let's compare the cell to a city.

Structure and function

Nucleus—The nucleus controls all the functions of the cell. In our city example, the nucleus would be the town hall. The town hall is where decisions are made in a city, and, similarly, the nucleus is where decisions are made in the cell. If the cell were your body, the nucleus would be the brain. The nucleus also contains most of the cell's DNA, which is used to make proteins for cell growth and reproduction.

The nucleus has a cell membrane surrounding it that is called the **nuclear membrane.** The nuclear membrane is double-layered and has many "holes" called pores that allow materials to go into and out of the nucleus. The nuclear membrane also functions to keep things out of the nucleus that could harm the genetic material inside. The nucleus contains the **nucleolus,** where ribosomes are made. (We will discuss ribosomes in more detail in chapter 2.)

Endoplasmic reticulum (ER)—The endoplasmic reticulum can have ribosomes attached, giving it a rough appearance (thus, it is called the "rough ER"), or it may not have ribosomes attached, in which case it is known as "smooth ER."

The smooth ER is where lipids for the cell membrane are made, and the rough ER is where proteins are produced (we will discuss lipids and proteins in the "Biomolecules" section of this chapter).

Ribosomes—Ribosomes are the parts of the cell that make proteins. Ribosomes can be free-floating in the cell or attached to the rough endoplasmic reticulum. Ribosomes would be a factory in our city. (Protein synthesis is covered in chapter 2.)

Golgi apparatus—If the ribosomes are the factory of our city, the Golgi apparatus would be the part of the factory that focuses on processing and shipment of goods. The Golgi apparatus processes, sorts, packages, and sends out proteins to their final destination. The Golgi apparatus might send them to other parts of the cell or outside the cell.

Lysosomes—Lysosomes would be both the police and the garbage pickup crew of the cell. The lysosomes help to rid the cell of broken-down or worn-out cell parts, while also helping to defend the cell from bacteria and viruses. Once the lysosomes break down the worn-out cell parts, they then put the pieces back into the cytoplasm for the cell to use again.

Cytoplasm—The cytoplasm is the liquid part of the cell from outside the nucleus to the cell membrane. The organelles are suspended within the cytoplasm. In our city example, the cytoplasm would be the air, but in the cell it is liquid. The cytoplasm could be thought of as a pool of water in which swimmers are floating.

Cytoskeleton—Organelles do not actually float freely about the cytoplasm. Instead, the cytoplasm has a network of microtubules and microfilaments that support and hold the organelles in place in the cell. Microtubules are long, hollow tubes that give the cell its shape; microfilaments are smaller proteins that help the cell to move and divide.

Vesicles—Vesicles transport materials from one place to another inside the cell. Think of the vesicles as the 18-wheeler semi trucks that transport goods from one place to another.

Vacuoles—Vacuoles store materials for use later or for disposal. Think of a vacuole as a storage unit in a cell. Just as you may rent a storage unit and go get things out when you need them, or store things there until you can sell or donate them, the vacuoles do this for the cell. In a plant cell, a special vacuole called the central vacuole is a large organelle that takes up most of the space in the cell. Water is stored in the central vacuole of a plant cell, and if the plant does not get enough water, the central vacuole will shrink and the plant will wilt. In animals, vacuoles are much smaller than the single large one found in plant cells.

Mitochondria—The mitochondria are the power plants of the cell. Just as a power plant provides energy to a city, the mitochondria provide energy to the cell. Mitochondria take the food you eat and convert it into energy. Mitochondria have a double membrane: The inner membrane is folded many times to increase the surface area, and the outer membrane is smooth.

Cell membrane—The outside of the cell contains a cell membrane that gives the cell protection and support and allows things to come into and out of the cell. The cell membrane is semipermeable, allowing only certain substances to pass through (more on this in the next section of this chapter, "Cellular Processes"). The cell membrane would be the chain-link fence around a city.

Cell wall—Cell walls are found in plants, fungi, algae, and most prokaryotes. Cell walls are rigid structures that provide support for the cell. Cell walls are permeable and allow things to enter and leave the cell. The cell wall would be a stronger fence that would surround a city.

Chloroplasts—Plants and other photosynthetic organisms contain chloroplasts. Chloroplasts contain chlorophyll, a molecule that captures sunlight to perform photosynthesis and also gives photosynthetic cells their green color. Chloroplasts would be like a restaurant in the city. A restaurant makes food for the people in the city, and the chloroplasts make food for the plant.

Plant cells have many of the same organelles as animal cells, but some things are different. For example, plant cells contain chloroplasts. Figure 1-3 shows a plant cell; compare it to the animal cell from the beginning of this section (Figure 1-2, page 8).

Figure 1-3 Plant cell.

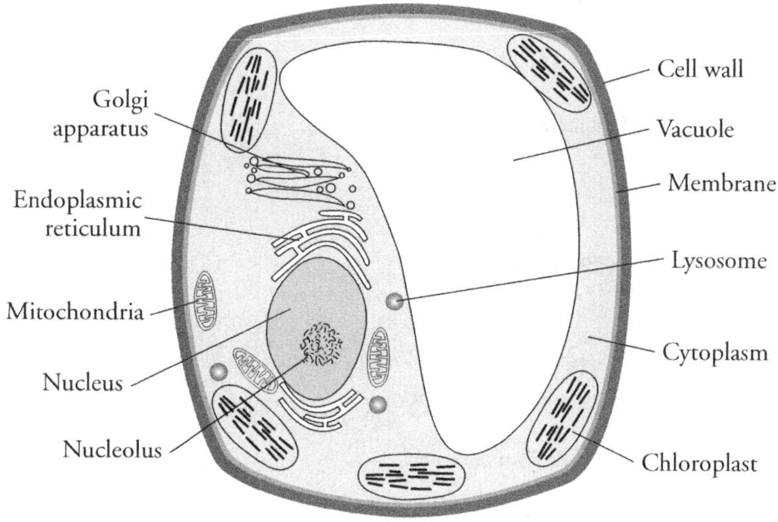

Cellular Processes

Now that we have discussed the structures and functions of a cell, let's look at the processes that cells go through to help them stay alive.

Homeostasis

Organisms must be able to maintain certain conditions for them to stay alive. **Homeostasis** is the term we use to describe the way organisms react to their internal and external environments to remain stable.

Now let's look at the different functions the cell performs in order to maintain homeostasis.

Cell transport

Think back to the cell membrane's function. We said that the cell membrane gives the cell protection and support and screens molecules that come into and out of the cell. This means it is semipermeable and is selective on what it allows in and out. Figure 1-4 shows a cell membrane.

Figure 1-4 Cell membrane.

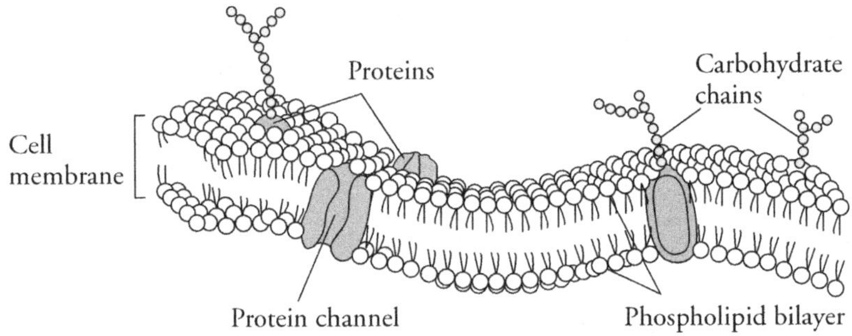

The picture looks complicated, but let's break it down into its parts and the function of each of those parts.

Can you tell from Figure 1-4 that the cell membrane is double layered? It is a double layer of phospholipids. A **phospholipid** is a molecule that is made up of three parts: a charged phosphate group, glycerol, and two fatty-acid chains. See Figure 1-5 for a more detailed look.

Figure 1-5 Phospholipid molecule.

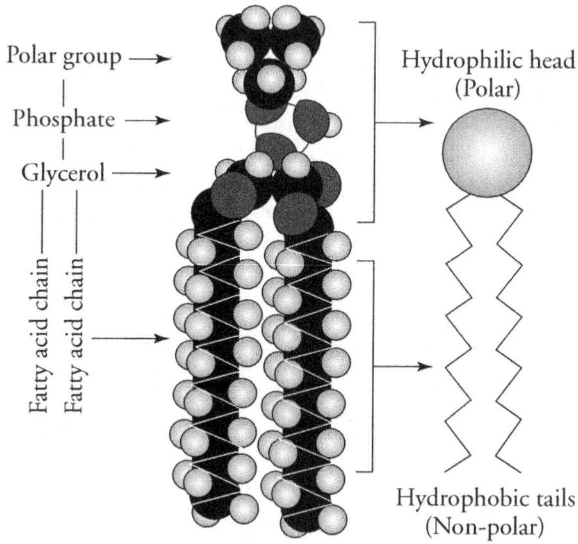

The phosphate and glycerol form the "head," and the fatty-acid chains form the "tails."

Can you also see from Figure 1-4 that there are proteins found in between some of the phospholipid bilayer? These proteins help the cell to move materials into and out of the cell.

The cell membrane is **selectively permeable,** meaning that in order for a cell to maintain homeostasis, it must allow some things in and out and prevent the movement of others, depending on the conditions at the time. The cell does this by diffusion and osmosis.

Passive transport

Diffusion is the movement of materials from a higher concentration to a lower concentration. An example would be if there were two rooms that were connected to each other by a door. There are 100 people in one room, and the other is empty. What do you think would happen if the door between the two rooms were opened? Some of the people would want a little more space and would begin moving to the empty room until there were about an equal number of people in each room.

Some substances do not easily diffuse across a membrane like water does. Such substances may need help by a process called **facilitated diffusion.** Facilitated diffusion uses transport proteins to help or "facilitate" the movement. This is still a form of passive transport and does not require any energy from the cell.

Osmosis is the diffusion of water across the cell membrane (see Figure 1-6). If there is more water inside the cell than outside the cell, it is in a **hypertonic solution** and water will move out of the cell.

A cell with equal amounts of water inside and outside is said to be in an **isotonic solution.** Water will continue to move into and out of the cell, but at the same rate.

Finally, when there is more water outside the cell than inside the cell, the cell is said to be in a **hypotonic solution.** Water is greater outside the cell, so the water will move into the cell.

Figure 1-6 Osmosis.

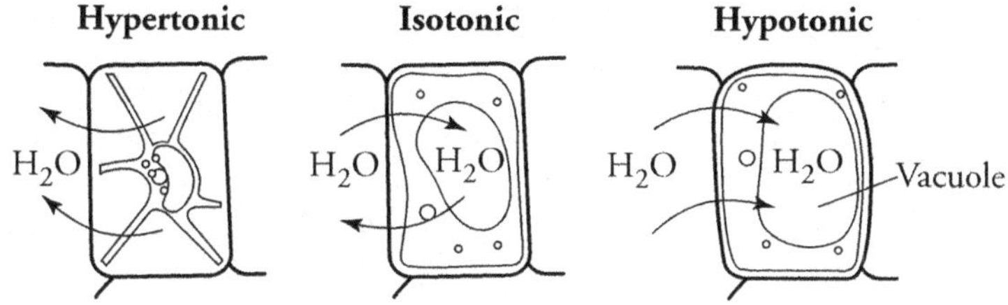

Active transport

Sometimes a cell needs to move substances from a lower concentration to a higher concentration. In this situation, diffusion and osmosis will not suffice. This is called **active transport.**

Moving substances against the gradient will take energy from the cell in the form of ATP. This type of movement is done by **carrier** or **transport proteins** that are found in the cell membrane. See Figure 1-7 for a comparison of passive vs. active transport.

Figure 1-7 Passive transport vs. active transport.

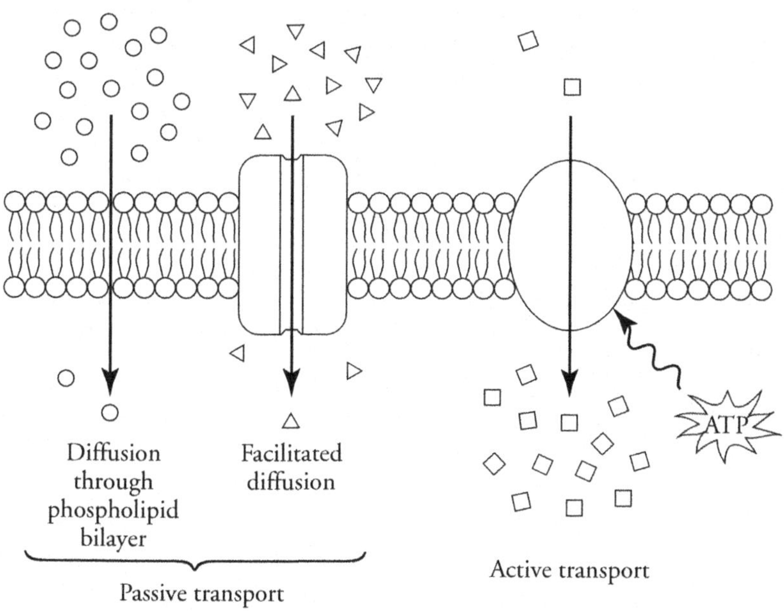

A cell might also need to move a large substance across the membrane that will not be able to move through the transport proteins. This is still active transport.

Endocytosis (Figure 1-8) is when a cell takes in large molecules by engulfing them. The cell membrane forms around the molecule and then pinches off to form a vacuole inside the cell. The lysosomes would then come and rid the cell of the membrane, leaving the molecules in the cell.

Figure 1-8 Endocytosis.

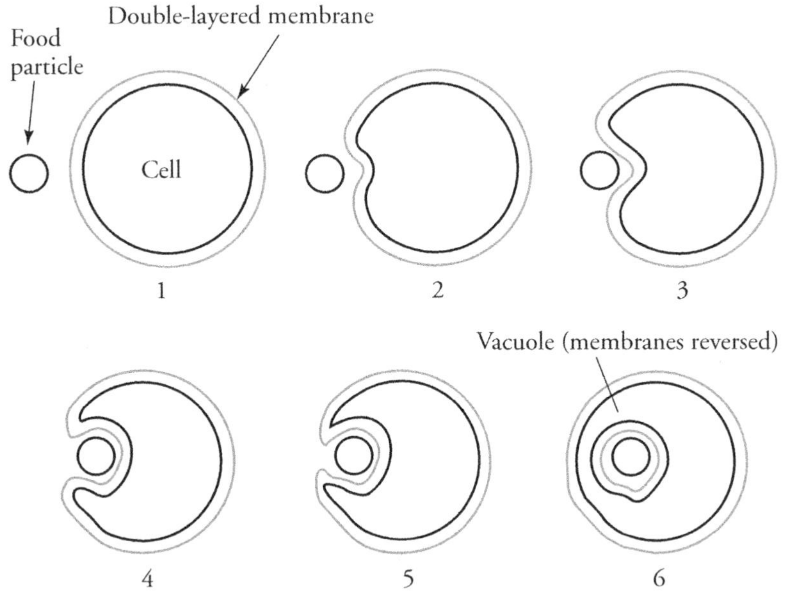

Exocytosis (Figure 1-9) is the opposite of endocytosis. Exocytosis is when the cell needs to move a large molecule from inside the cell to outside the cell, and transport proteins will not work.

Figure 1-9 Exocytosis.

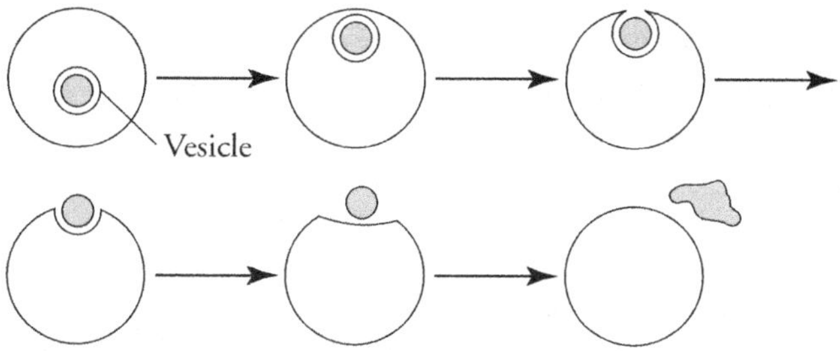

Viruses

In 1892, a Russian scientist by the name of Dmitri Ivanovsky noticed that something was infecting the tobacco plants that he was studying. The leaves of the tobacco plants were being scarred. Most people thought that the thing attacking the tobacco plants was a bacterium. However, when Ivanovsky strained the substance through a filter that had holes too small for bacteria to fit through and then put the strained liquid onto the leaves of the plant, the plant became infected and the leaves scarred.

Then, in 1897, Dutch scientist Martinus Beijerinck continued Ivanovsky's work. Beijerinck suggested that tiny particles in the liquid were causing the infection. He called these particles **viruses,** from the Latin word for "poison."

It wasn't until 1935 that an American scientist named Wendell Stanley discovered that the tobacco mosaic virus, as it became known, was able to crystalize. Since living organisms do not crystalize, he decided that viruses were therefore not living.

We now know that viruses are able to enter living cells and, by using the cell's molecular machinery, produce more viruses. Some viruses that you have probably heard about are the human immunodeficiency virus (HIV) and influenza (flu).

Virus structure

A virus is a non-living biological entity composed of genetic material within a protein coat. Viruses are extremely small and can only be seen using an electron microscope.

A single virus is called a **virion,** and all viruses have an outer layer called a **capsid.** The capsid is made of proteins, and inside the capsid the genetic material is found. The proteins on the outer surface of the capsid help the virus to enter the host cell. Some viruses have spikes on the outside that help them to attach to the host cell. Some viruses have DNA as their genetic material, and others have RNA, but no viruses have both DNA and RNA. Viruses have different forms depending on the type of virus (see Figure 1-10).

Figure 1-10 Virus forms.

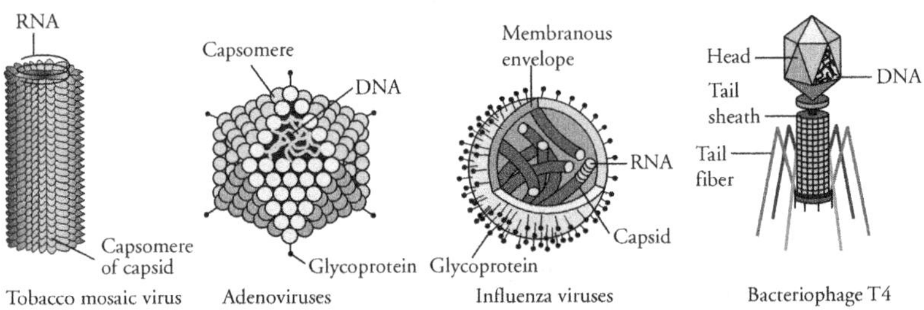

Viral reproduction

Viruses are not alive and cannot reproduce on their own. A virus must invade a host cell and get the host cell to do its dirty work. The virus attaches to the outside of the host and injects viral genetic material into the host. The host then reproduces the virus.

There are two ways that a virus can reproduce; one is known as the lytic cycle and the other as the lysogenic cycle.

In the **lytic cycle** (Figure 1-11), the virus enters the host cell, has the host cell make copies of the viral genetic material, and then causes the host cell to burst open. Once the host cell bursts, hundreds of viruses are released that go infect more cells.

Figure 1-11 Lytic cycle.

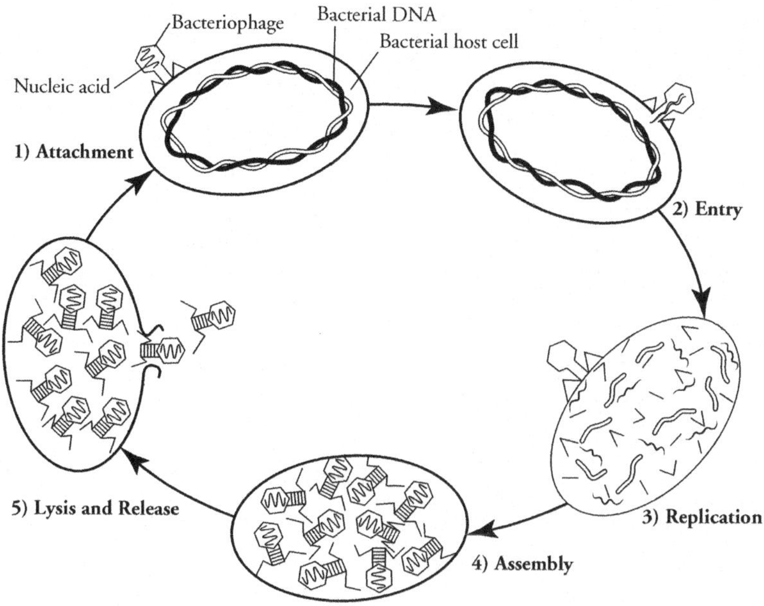

The second mode of infection is the **lysogenic cycle** (Figure 1-12), in which the virus enters the cell and actually causes the host cell's DNA to combine with viral DNA. When the viral DNA is integrated into the host's genome, it is called a **prophage.** The host cell does not know that its DNA has been altered and continues through the cell cycle, making copies of the altered DNA.

Figure 1-12 Lysogenic cycle.

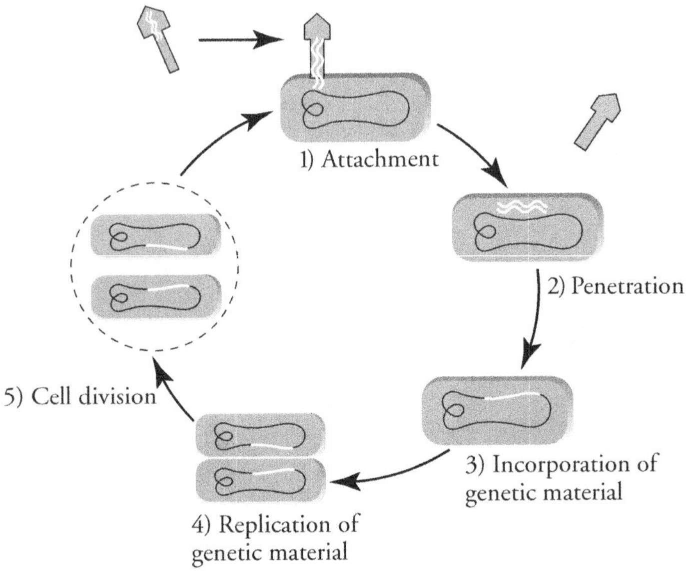

The virus can stay dormant for years before becoming active. Then, due to stress, illness, or other factors, the virus can activate, remove its DNA from the host DNA, and cause the host to begin making copies and producing new viruses. At this point, the cell leaves the lysogenic stage and begins the lytic stage.

Retroviruses

Some viruses have RNA instead of DNA. These **retroviruses** get their name because they go backward (*retro* means "backward"). So instead of going from DNA to RNA, they go from RNA to DNA. A retrovirus will make DNA from RNA and then combine with the host cell's DNA, where it begins a lysogenic cycle and can remain dormant for many years. Once it activates, it goes into the lytic cycle, making new viruses.

Cell Cycle and Mitosis

The cell cycle (Figure 1-13) is the way in which all organisms grow, duplicate their DNA, and divide to create two identical cells.

Figure 1-13 Cell cycle.

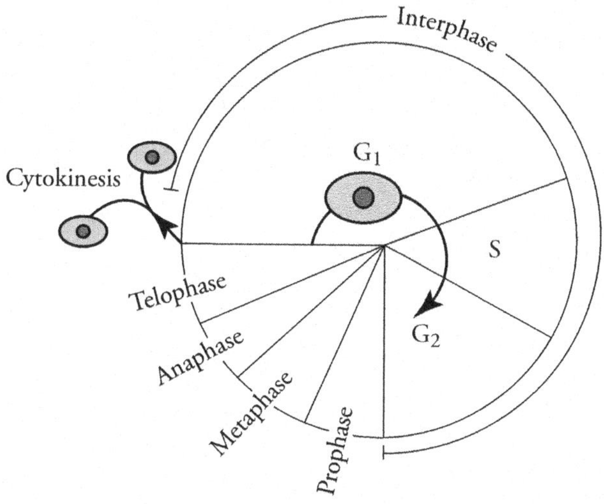

Cell cycle

In order for your body to grow and repair itself, your cells must be able to divide. The growth and division of cells is called the **cell cycle.** During the cell cycle, cells grow, divide, form two identical cells, and then repeat the cycle all over again. The cell cycle has four stages: gap 1 (G_1), synthesis (S), gap 2 (G_2), and mitosis. The average time it takes for a human cell to divide is 12 to 24 hours, but that depends on what kind of cell it is. Also, children's cells divide faster and, as we age, the cycle slows down and our cells divide slower.

Interphase

The longest part of the cell cycle is **interphase,** which includes the G_1, S, and G_2 phases. During interphase, the cell may be performing its normal cell functions—growing, making proteins, and copying its DNA—depending on the requirements of the cell. Let's break down interphase into what happens at each of the three phases.

During $\mathbf{G_1}$, the cell grows, replicates its organelles, and performs its normal functions. If it is a skin cell, it protects; if it is a heart cell, it contracts heart muscle; and so on. Most of the cell's time is spent in the G_1 phase. During this phase, the DNA is not tightly coiled and is referred to as **chromatin.**

During the **S** phase, a cell is replicating its chromosomes and making a copy of its DNA. At the end of the S phase, the cell will have two identical copies of its DNA. A cell will only move from the G_1 phase into the S phase if it receives chemical messages telling it to divide.

Once the cell has two identical copies of its DNA, it enters the **G_2** phase, where the cell continues to grow and perform its normal activities. In addition, certain organelles needed for cell division are made.

Through the G_1, S, and G_2 phases, the cell has just become prepared for cell division. The cell actually divides during a process called mitosis.

Mitosis

Mitosis is the fourth stage of the cell cycle, where the nucleus will divide into two nuclei and then the rest of the cell will divide, forming two identical daughter cells. Mitosis is further broken up into four phases: prophase, metaphase, anaphase, and telophase. When the cytoplasm divides in half, it is called cytokinesis. Let's look at the four phases.

Prophase

The longest phase of mitosis is **prophase** (Figure 1-14), where the cell's chromosomes become visible and the centrioles move to opposite sides of the nucleus. The **centrioles** are the structures that will pull the chromosomes apart during anaphase. Specialized structures called spindle fibers will begin to form, radiating outward from the centrioles and stretching to the chromosomes. When the cell's chromatin begins to tighten and coil, it becomes a **chromosome.**

Figure 1-14 Prophase.

Metaphase

The second phase is **metaphase** (Figure 1-15), where the **spindle fibers** attach to the chromosomes and the chromosomes line up in the middle of the nucleus. The spindle fibers assist the centrioles during anaphase to separate the chromosomes. It is important during metaphase that the chromosomes line up perfectly so that once anaphase begins, they will divide equally.

Figure 1-15 Metaphase.

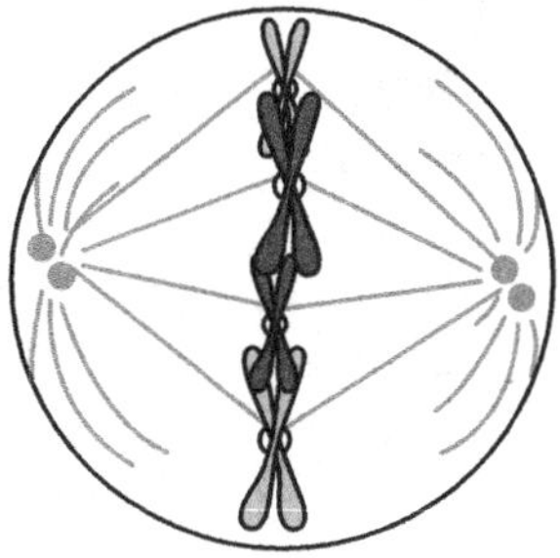

Anaphase

The next phase is **anaphase** (Figure 1-16), where the centrioles and spindle fibers pull the sister chromatids apart. **Sister chromatids** are the identical chromosomes that form in the nucleus of each new cell. At the end of anaphase, the chromosomes will be at opposite ends of the nucleus.

Figure 1-16 Anaphase.

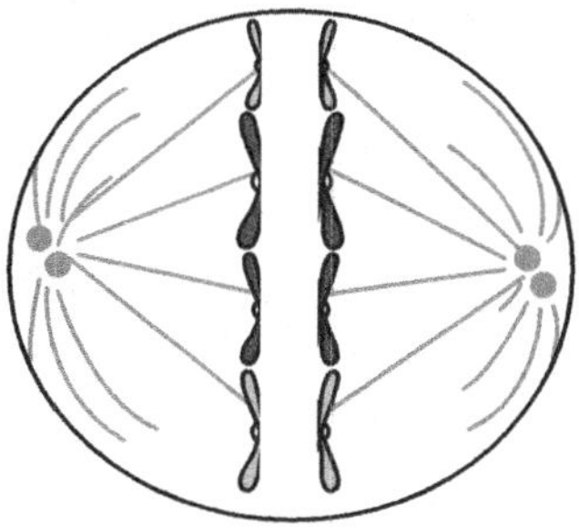

Telophase

The final phase of mitosis is **telophase** (Figure 1-17). During telophase, the chromosomes begin to unwind and relax again, forming a stringlike mass; a nuclear membrane begins to form around the DNA strands, and the nucleolus appears in each daughter nucleus.

Figure 1-17 Telophase.

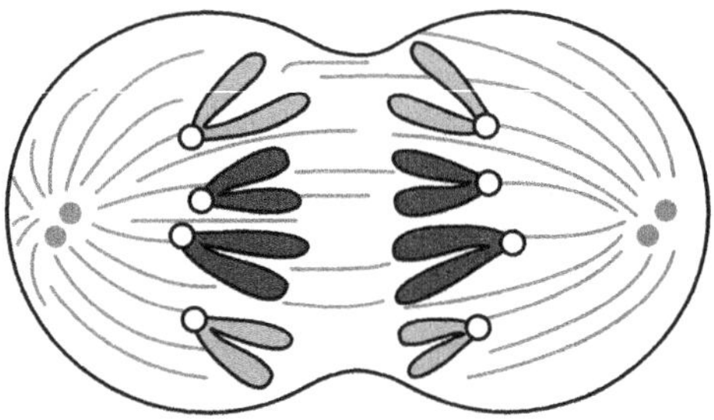

Cytokinesis

After mitosis, cell division is still not complete. We now have two identical nuclei, but we do not have two identical cells. For cell division to be complete, cytokinesis (Figure 1-18) must occur. **Cytokinesis** is division of the cytoplasm; it usually begins during telophase. What results are two cells that are identical; we now begin the G_1 phase all over again. This cycle continues your entire life.

Figure 1-18 Cytokinesis.

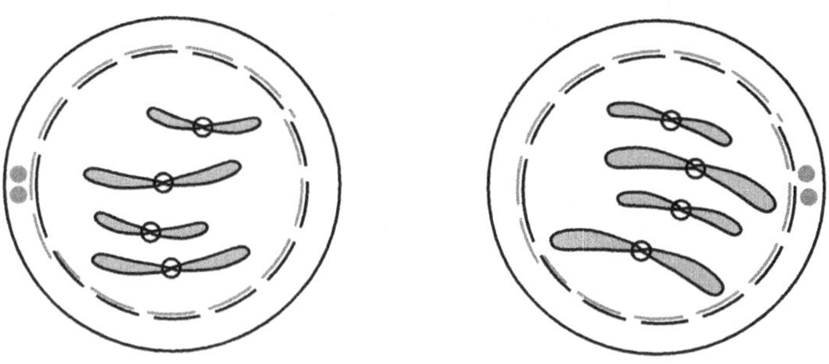

Cell differentiation

As discussed earlier, the cells in your body perform different functions. The different jobs that your cells do are referred to as **cell differentiation.** We said your heart cells help the heart to beat and your skin cells protect you. Plant cells have different jobs as well. The plant has specialized cells for the roots to absorb water and specialized cells in the leaf to perform photosynthesis. Different cells also divide at different rates. The cells of your nervous system divide extremely slowly, while the cells in your stomach divide very quickly. The time it takes for each cell to divide depends on the job that it performs.

Cancer

So, what happens when cells do not divide properly? If cell division does not happen precisely, you may get mistakes. Most of the time, nothing happens, and you would never know that your cells divided (or attempted to divide) incorrectly. However, due to your family's genetics, chemicals in the environment, and mistakes that can occur during the cell cycle, your cells might not get away with this "bad" cell division. When a cell divides incorrectly, it might become a cancer cell.

The problem with cancer cells is that they divide very quickly and can form masses called tumors. Cancer cells do not spend much time in interphase like normal cells do. This allows the cancer cells to grow faster than normal cells and can cause very serious problems or even death.

Chemicals that have been proven to promote cancer cell growth are called **carcinogens.** However, not all cancer is due to carcinogens or a family history of cancer. Sometimes there are changes or mutations in a cell's DNA that can cause cancer. Other factors like prolonged exposure to UV light can stimulate a cell to mutate, which can lead to cancer.

Biomolecules

All living things are made up of carbon; therefore, **carbon** is thought of as the building block of life. In fact, organic chemistry is an entire branch of science devoted to just studying carbon. Why is carbon so special?

A carbon atom (Figure 1-19) has four valence electrons in its outer shell, allowing carbon to bond with other elements like hydrogen, oxygen, nitrogen, phosphorus, sulfur, and other carbon atoms. The type of bond that carbon makes with other elements is a covalent bond; because of this characteristic of carbon, it can form millions of different structures, leading to the diversity of life on our planet. A **covalent bond** is when there is a sharing of electron pairs between atoms.

Figure 1-19 Carbon atom.

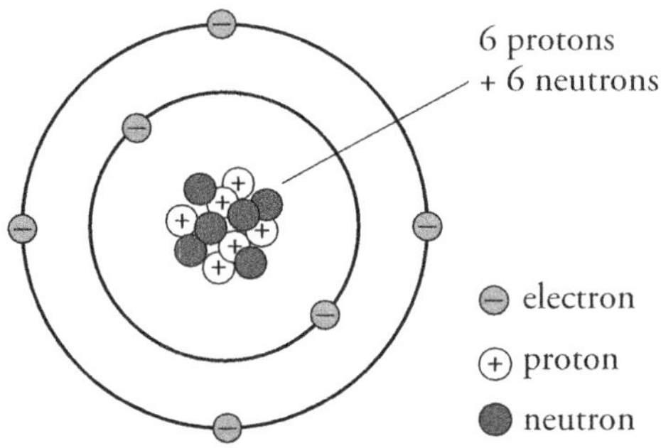

A **macromolecule** is a large molecule. A macromolecule is considered a **polymer** if it is composed of repeating smaller molecules called **monomers.** We will now look at the four most important macromolecules in living things: proteins, carbohydrates, lipids, and nucleic acids.

Proteins

A **protein** is a polymer made up of monomers called amino acids. **Amino acids** are compounds where carbon is bonded with a hydrogen atom, an amino group ($-NH_2$), and a carboxyl group ($-COOH$). In addition, each amino acid will have a variable group ($-R$), which is what makes each amino acid different. We have 20 amino acids that make protein, and the variable group is what makes each amino acid dissimilar. Figure 1-20 shows the structure of an amino acid.

Figure 1-20 Amino acid structure.

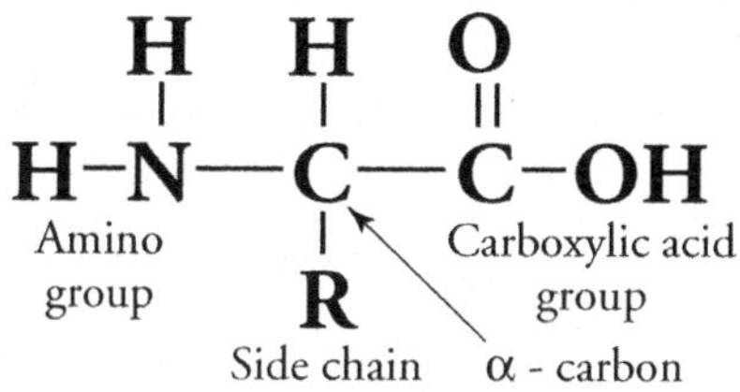

Amino acids link together in chains called **polypeptides.** One or more polypeptides make a protein. Proteins are very important to living things; almost all functions that occur in your body involve protein. For example, proteins are important in bone and muscle development, your skin and hair are made of protein, and proteins make up the hemoglobin that transports oxygen to your cells.

Carbohydrates

Carbohydrates are made up of carbon, hydrogen, and oxygen. Most carbohydrates are found in a 1:2:1 ratio; for example, glucose is $C_6H_{12}O_6$. Carbohydrates are an important source of energy for living things.

The three types of carbohydrates are monosaccharides, disaccharides, and polysaccharides. A **monosaccharide** is a simple sugar like glucose, galactose (found in milk), and fructose (found in fruit). Two monosaccharides can combine to form a **disaccharide.** Sucrose (table sugar) and lactose (also found in milk) are examples of disaccharides. Finally, when more than two monosaccharides are combined, we get a **polysaccharide.** Starches (energy storage for plants), cellulose (the structural component of plant cell walls), and glycogen (found in your liver and skeletal muscle) are all polysaccharides composed of the monomer glucose. Figure 1-21 shows the structure of the monosaccharides glucose and fructose.

Figure 1-21 Monosaccharides.

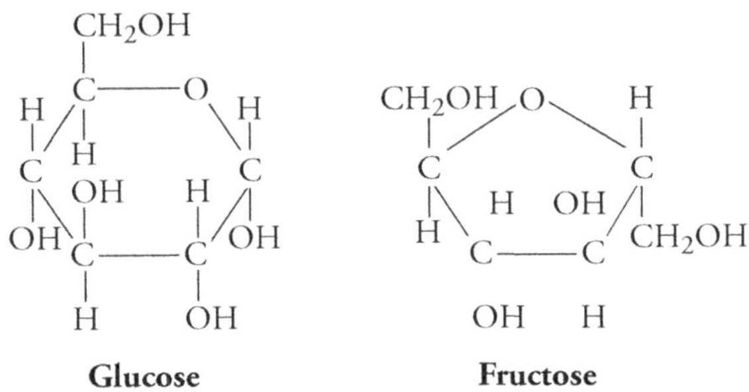

Lipids

Lipids are important to living things because they store energy, act as chemical signals, and are important components of cell membranes. Fats, oils, wax, and cholesterol are common types of lipids. Specific types of lipids called **fats** are made up of three fatty-acid molecules joined with a glycerol. Fats can either be solid or liquid at room temperature, depending on the chemical nature of their fatty acids.

Have you heard different food commercials talk about saturated, unsaturated, and polyunsaturated fats? These terms refer to the number of double carbon–carbon bonds found within a fat molecule's fatty-acid chains. A **saturated fat** has only single carbon–carbon bonds within the fatty-acid chains. This results in two hydrogen atoms bonding to each carbon atom (since each carbon atom will participate in a total of four covalent bonds). An **unsaturated fat** (Figure 1-22) has a double bond between carbon atoms so there are fewer hydrogen atoms attached to each carbon. A **polyunsaturated fat** would have more than one of these double bonds of carbon atoms; therefore, even fewer hydrogen atoms are found.

Earlier in this chapter, we learned about cell membranes and how they contain a double layer of phospholipids. A **phospholipid** is another type of lipid that is important in living things.

Figure 1-22 Unsaturated fat.

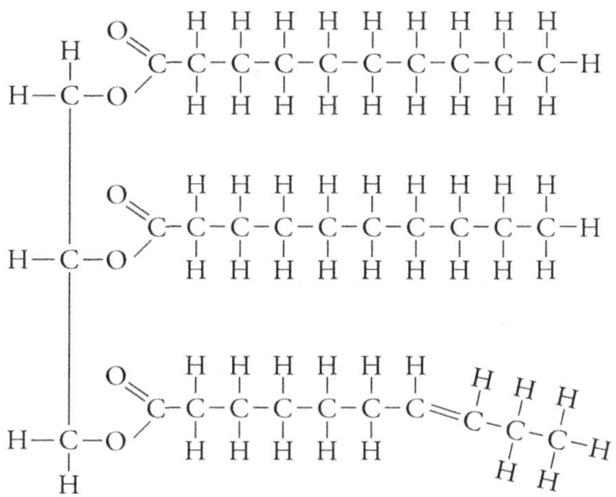

Nucleic acids

The fourth and final macromolecule that we will discuss are nucleic acids. **Nucleic acids** store and transmit genetic information. The two types of nucleic acids are deoxyribonucleic acid (DNA) and ribonucleic acid (RNA). A nucleic acid is made up of smaller units called nucleotides. **Nucleotides** all have a phosphate, a ribose sugar, and a nitrogenous base. Figure 1-23 depicts a nucleotide.

Figure 1-23 Nucleotide.

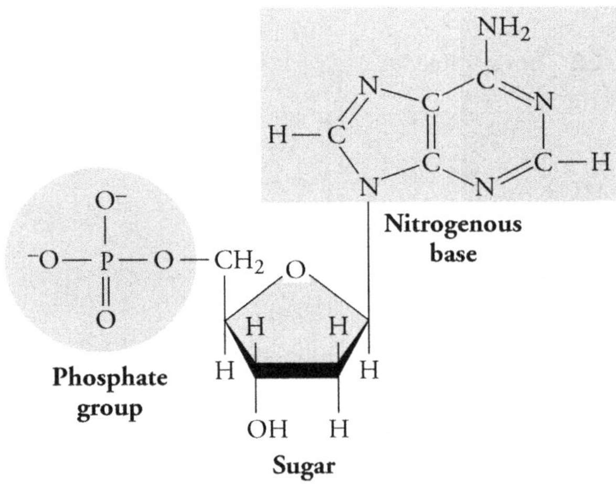

Chapter Check-Out

1. You are a researcher trying to discover why a group of passengers on a cruise ship have become sick. You take a sample of an ill passenger's blood and look at the sample under an electron microscope. The picture below shows what you see under the microscope.

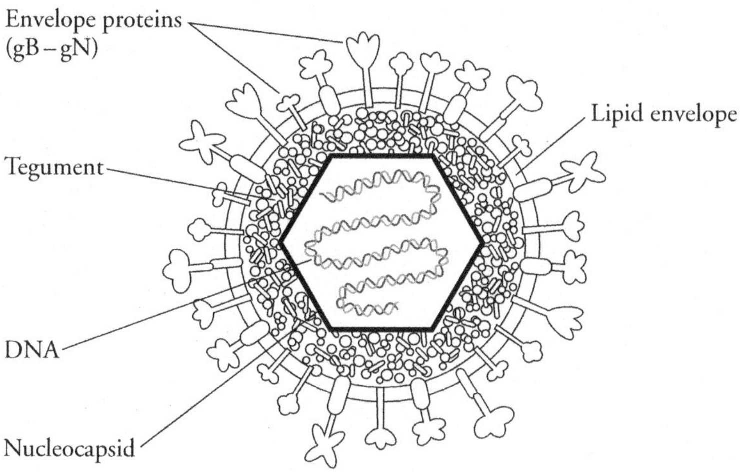

   You conclude that the passengers have been infected by a(n)

   a. bacteria.
   b. bacteriophage.
   c. influenza virus.
   d. protozoan.

2. Which of the following occurs during the G_1 phase of the cell cycle?

   a. cell growth
   b. DNA replication
   c. preparation for mitosis
   d. mitosis

3. Which stage of mitosis is shown in Step 4 below?

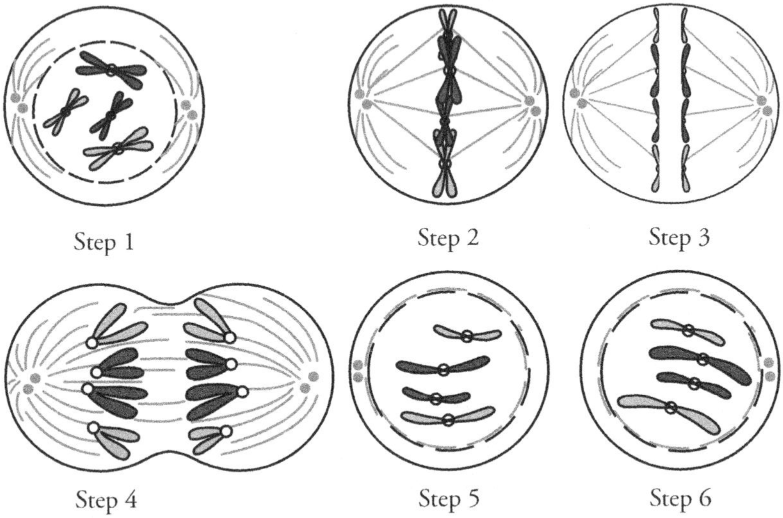

  a. interphase
  b. anaphase
  c. telophase
  d. metaphase

4. A cell membrane has a double layer that protects the cell. This double layer is made up of a type of macromolecule. Which macromolecule makes up the cell membrane?

  a. carbohydrate
  b. protein
  c. amino acid
  d. lipid

5. Which of the following cellular structures would be present in a eukaryotic cell and not a prokaryotic cell?

  a. cell membrane
  b. genetic material
  c. nuclear membrane
  d. cytoplasm

6. Your body is able to take the food you eat and transform it into chemical energy. Which of the following cellular organelles would perform this function?
   a. mitochondrion
   b. ribosomes
   c. endoplasmic reticulum
   d. nucleus

7. Which stage of viral reproduction is pictured below?

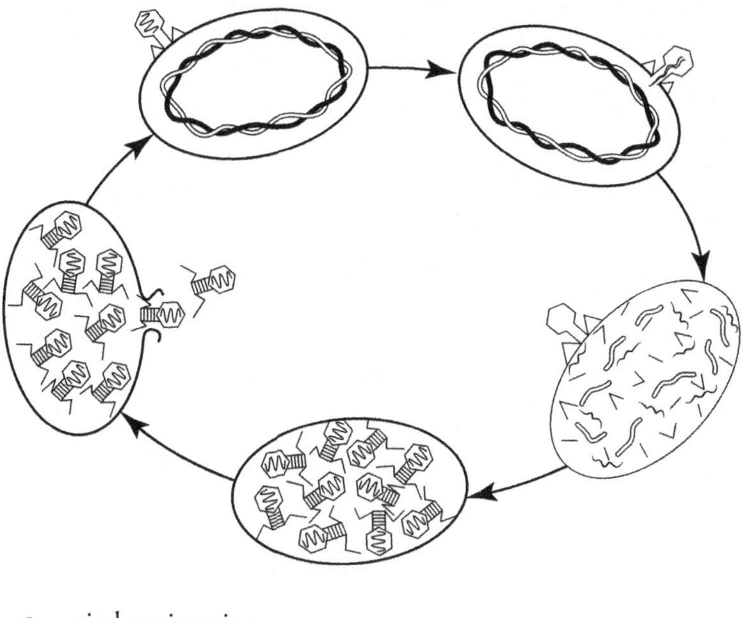

   a. viral conjugation
   b. lytic cycle
   c. lysogenic cycle
   d. facilitated diffusion

8. During the cell cycle, the cell is able to grow, replicate, and divide. However, if a cell develops cancer, a mass of cells called a tumor develops. What has happened to cause the cell to become cancerous?
   a. The cell is unable to perform mitosis.
   b. The cell stays in the S phase and cannot move to the next phase.
   c. The cell cycle continues without regulation.
   d. The cell is unable to continue in the G_1 phase.

9. White blood cells in the human body are responsible for helping to rid the body of harmful substances and cells that have outlived their usefulness. For example, a white blood cell might surround a damaged red blood cell and engulf it, bringing the red blood cell into the white blood cell. This is an example of

a. endocytosis.
b. exocytosis.
c. osmosis.
d. passive transport.

10. You take a trip down to South Padre Island and float in the swimming pool for hours. You get out of the water, and your fingers are all puffy and swollen. This has occurred because you were in a(n)

a. hypotonic solution.
b. hypertonic solution.
c. isotonic solution.
d. pinocytosis solution.

Answers: 1. c **2.** a **3.** c **4.** d **5.** c **6.** a **7.** b **8.** c **9.** a **10.** a

Chapter 2
MECHANISMS OF GENETICS

Chapter Check-In

- ❑ DNA and RNA (replication, transcription, and translation)
- ❑ Genetic crosses
- ❑ Meiosis

In 1928, British scientist Frederick Griffith was studying pneumonia in mice and made a very interesting discovery. Griffith was studying two *strains* (types) of pneumonia: a smooth strain and a rough strain. He injected the smooth strain that caused disease into a mouse. The mouse died of pneumonia. Then he injected the harmless (rough) strain into a mouse, and the mouse lived. Next, he heated the smooth, disease-causing strain to kill the bacteria. He injected it into the mouse, and the mouse lived. Finally, he injected a mouse with a mixture of the heat-killed disease-causing strain and the living, harmless strain, and the mouse died. Neither of these two inoculations, if given alone, would kill the mouse. Why did the mouse die if the inoculations were injected as a mixture? Furthermore, why was Griffith able to isolate from the dead mouse the living smooth strain (which was not injected into the mouse in the first place)? He concluded that the living harmless form had taken on some of the characteristics of the heat-killed, disease-causing form and "transformed." He called the process **transformation** (Figure 2-1). This experiment started the search for this transforming substance that we now know as DNA.

Figure 2-1 Transformation.

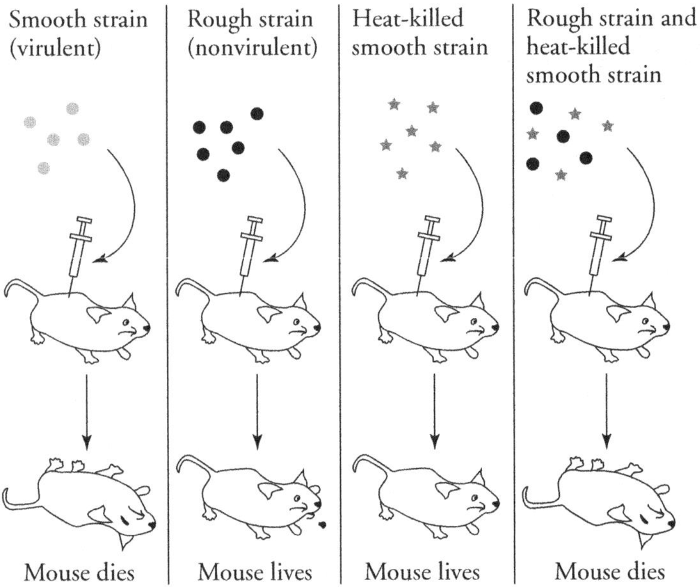

DNA and RNA: Replication, Transcription, and Translation

DNA structure

Deoxyribonucleic acid, or **DNA,** is a polymer made up of nucleotides, each composed of a five-carbon sugar, a phosphate group, and a nitrogenous base (see Figure 2-2). In DNA, there are four types of nitrogenous bases: cytosine (C), thymine (T), adenine (A), and guanine (G). Cytosine always joins with guanine, and thymine always joins with adenine. One way to remember this is Good Cats Always Together (think "Good Cats," so the G and C are joined, and "Always Together," so the A and T are joined). These are the rules of base pairing. DNA is also spiral-shaped and is referred to as a double helix. Figure 2-2 is a drawing of DNA.

Figure 2-2 DNA.

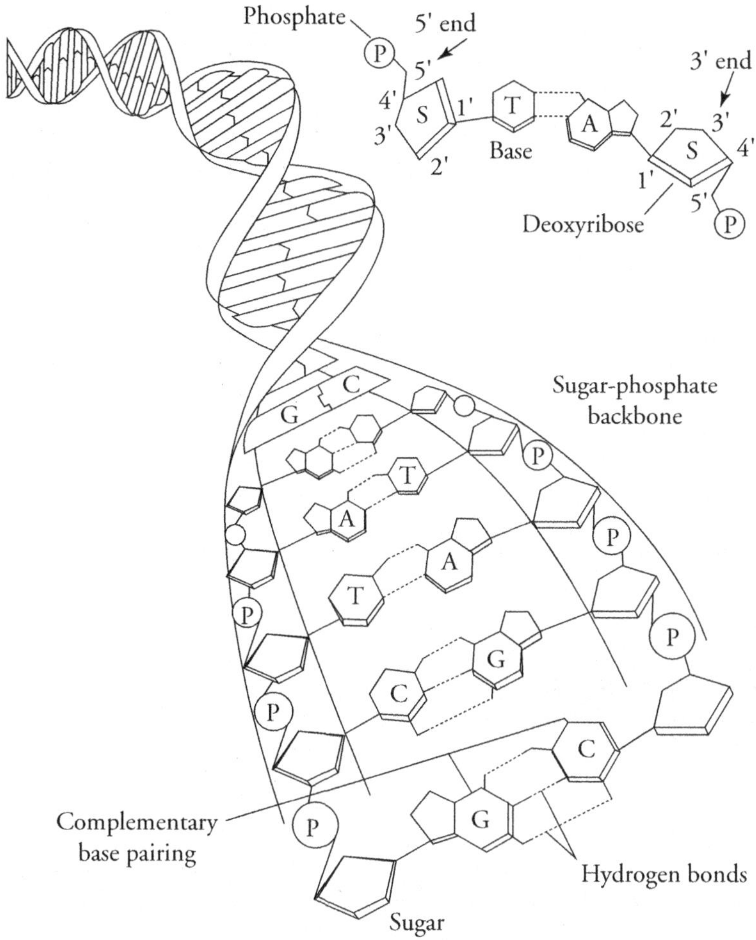

If you look closely at the picture of DNA, the backbone of DNA is formed by the sugar and phosphates. If you remember from chapter 1, nucleic acids are made up of smaller units called nucleotides, and each nucleotide has a phosphate, a sugar, and a nitrogenous base. One nucleotide will join with another, each with its own nitrogenous base (adenine, guanine, and so on). The sugar of one nucleotide is covalently bonded with the phosphate of the next one, with the nitrogen bases joining in the middle. DNA is often compared to a twisted ladder, with the sugars and phosphates making up the rails of the ladder and the bases making up the rungs of the ladder and then twisting it into a spiral shape.

It is also important to know that the nitrogenous bases are joined together by weak hydrogen bonds. This will be important in the next section of this chapter when we learn how DNA is able to replicate itself. When paired together, adenine and thymine form two hydrogen bonds, whereas guanine and cytosine form three.

How did scientists first discover the shape of DNA? In the early 1950s, a scientist named Rosalind Franklin was studying DNA in England. She was using a technique called X-ray diffraction to find information about the structure by aiming an X-ray beam at samples of DNA. Her images suggested that DNA was a double-helix shape. At about the same time, two other scientists—Francis Crick from England and James Watson from the United States—were also looking at the structure of DNA using Franklin's X-ray images, as well as data from other researchers. They built a three-dimensional model out of metal and wood, connecting the bases together like puzzle pieces. Watson and Crick published their model in the journal *Nature.*

DNA replication

We learned in chapter 1 that before a cell divides, its DNA must first be copied. Watson and Crick suggested that the double-helix shape of DNA makes this possible. DNA will unwind from the tight coil of how it is found in the chromosome. It will then unzip, much like unzipping a zipper into its two separate parts. If you recall, adenine will always join with thymine and cytosine will always bond with guanine, so if the DNA unzips, an exact copy can be made. An enzyme named **helicase** is responsible for unwinding and unzipping the DNA, and the weak hydrogen bond between the bases makes this possible. Then free-floating nucleotides that are found in the nucleus can pair with the unzipped DNA strands; another enzyme, **DNA polymerase,** bonds the new nucleotides together. Because each side or strand of the DNA can make the other strand, the strands are said to be complementary. This process of copying DNA is called **replication.** DNA replication occurs during the S phase of the cell cycle and guarantees that each new cell has an exact copy and complete set of its DNA. Figure 2-3 illustrates the process.

Figure 2-3 DNA replication.

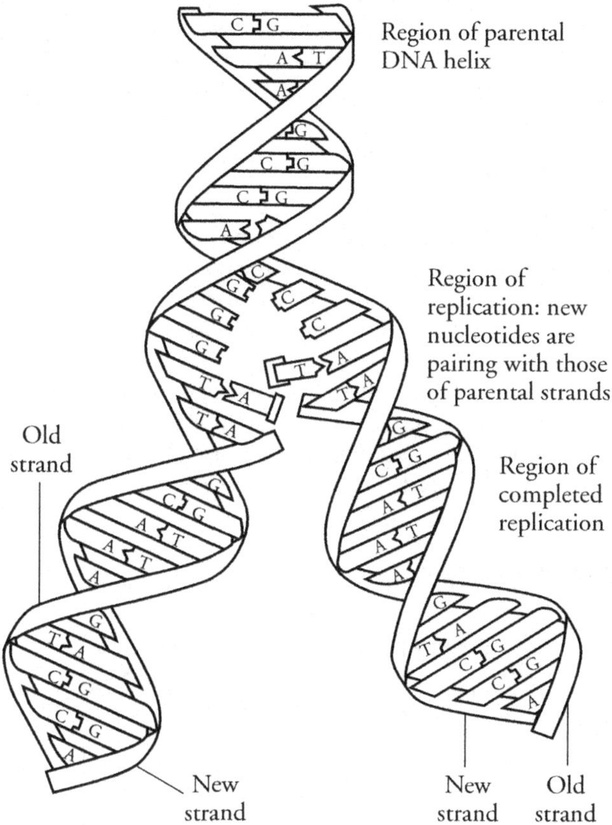

Note that each of the two new DNA strands is identical, and each has one new strand and one old strand. DNA replication occurs in hundreds of different places in the DNA of eukaryotes and happens in both directions.

RNA

Now that you understand how DNA replication occurs, we can discuss another important nucleic acid: **ribonucleic acid,** or **RNA.** You can think of RNA as a disposable copy of DNA. RNA is similar to DNA in that it has nucleotides with a sugar, a phosphate, and a nitrogenous base;

however, there are three important differences. First, the sugar in DNA is deoxyribose, but the sugar in RNA is ribose. Second, the nitrogenous bases in DNA and RNA all have guanine, cytosine, and adenine; but instead of thymine, RNA has a base called uracil. Uracil in RNA will bond with adenine in DNA. Finally, DNA is double-stranded, and RNA is a single strand.

Protein synthesis

Our cells need protein in order to grow and function, and the main purpose of RNA is **protein synthesis.** Protein synthesis (see Figure 2-4), or the putting together of amino acids to form protein, occurs in the ribosomes. The ribosomes are found in the cytoplasm of the cell, not in the nucleus. The cell does not want to risk allowing the DNA out of the nucleus, so a complementary RNA "copy" of certain segments of the DNA is created. This RNA serves as a "message" so the instructions for a protein can move from the nucleus, through the nuclear pores, and into the cytoplasm where the ribosomes are waiting. DNA is vital so the RNA, which is disposable, is used, and then is destroyed.

Figure 2-4 Protein synthesis.

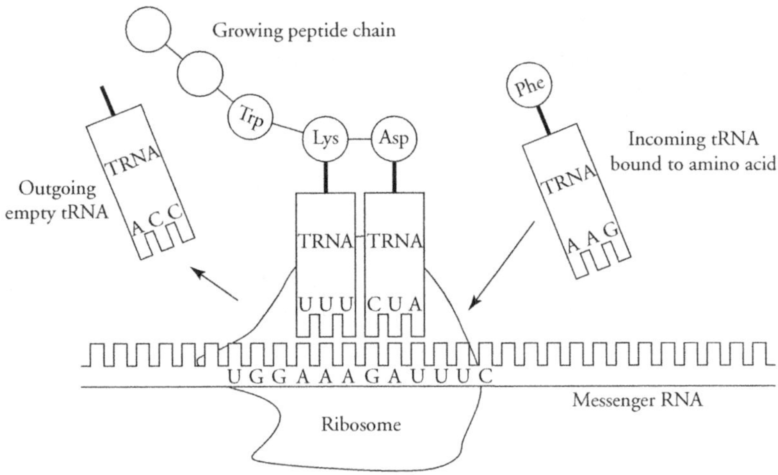

Transcription

Similar to DNA, transcription requires an enzyme to "unzip" the DNA. The enzyme used is RNA polymerase. **Transcription** is the process of taking a certain segment of the DNA, called a "gene," and making a complementary RNA copy. Just as in DNA replication, nucleotides string together to make a chain of RNA. The only difference here is that if the DNA nucleotide is adenine, the RNA base pair will be uracil, not thymine. Three types of RNA are used in translation: messenger RNA, transfer RNA, and ribosomal RNA. Let's look at each of these and the roles they play.

Messenger RNA, or mRNA (Figure 2-5), is like a waitress taking your order in a restaurant. You look at the menu, decide what you want to order, and give your request to the waitress. The waitress then takes your order to the cook to prepare your food. Similarly, mRNA takes the copy of the RNA from the nucleus and into the cytoplasm to the ribosomes. This RNA copy is the message that the ribosomes need in order to make the correct protein based on the specific gene that was transcribed. Remember, ribosomes are the protein factories of our cells.

Figure 2-5 Messenger RNA (mRNA).

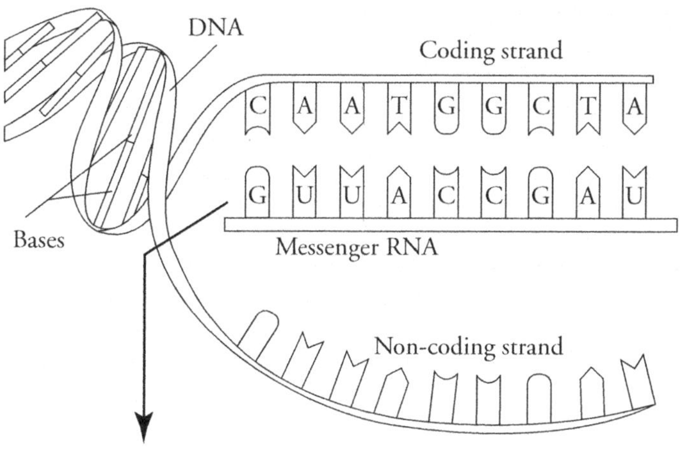

Translation

Once the mRNA takes the message to the ribosomes, it needs a way to translate that message into something the ribosome can understand. This is the process known as **translation.** It's as if the cook in the restaurant doesn't speak the same language as the waitress and needs someone there to translate the message from one language to another. This leads us to our second type of RNA: **transfer RNA,** or tRNA, which "translates" the message into polypeptides that make up a protein. Proteins are made up of 20 different amino acids. By using the four bases—adenine (A), guanine (G), cytosine (C), and uracil (U)—scrambled in different ways, tRNA can easily make up 20 different "codes" or amino acids.

A **codon** is a set of three nucleotides that code for a specific amino acid. Now think of the ribosomes as a factory. The mRNA is brought to the ribosomes, and the nucleotides are "read" three at a time. So, let's say that the first three nucleotides (or codon) are UUC. Only one specific type of tRNA will pair with the codon UUC and bring the correct amino acid. This pairing occurs between the tRNA's **anticodon** (AAG) and the mRNA's codon (UUC). Now the tRNA has "translated" the mRNA's message into an amino acid. The **ribosomal RNA** (rRNA) is responsible for holding the proteins in place and helping to join the amino acids together. This continues, binding the amino acids into a chain, until it reaches something called a "stop codon," which essentially tells the tRNA that the protein has been made and the chain breaks off.

Genetic code

Each codon, or set of three nucleotides, makes up a single amino acid. Every organism on this planet has a **genetic code.** For example, the codon UUG codes for the amino acid leucine. It doesn't matter if you are a human, a horse, or a hornet, the amino acid UUG is leucine. This also means that an amino acid from one organism can be inserted into another organism to make a useful protein. Scientists use techniques such as DNA fingerprinting, genetic modifications, chromosome painting, and chromosomal analysis to study the genes of different species.

Codon chart

Below are two types of codon charts (Figures 2-6 and 2-7). A **codon chart** allows you to find which amino acid is being coded for. Because RNA has four bases (G, C, A, U), there are 64 different three-base codons that can be formed, which is more than enough to code for the 20 different amino acids. There are also **stop codons** and **start codons** that signal to the ribosome to start making the protein and when to stop making the protein. Many codon sequences code for the same amino acid. For example, the codons CGU, CGC, CGA, CGG, AGA, and AGG all code for arginine; while only one codon sequence, AUG, codes for methionine.

Figure 2-6 RNA codon chart.

First letter	Second letter				Third letter
	U	**C**	**A**	**G**	
U	phenylalanine	serine	tyrosine	cysteine	**U**
	phenylalanine	serine	tyrosine	cysteine	**C**
	leucine	serine	stop	stop	**A**
	leucine	serine	stop	tryptophan	**G**
C	leucine	proline	histidine	arginine	**U**
	leucine	proline	histidine	arginine	**C**
	leucine	proline	glutamine	arginine	**A**
	leucine	proline	glutamine	arginine	**G**
A	isoleucine	threonine	asparagine	serine	**U**
	isoleucine	threonine	asparagine	serine	**C**
	isoleucine	threonine	lysine	arginine	**A**
	(start) methionine	threonine	lysine	arginine	**G**
G	valine	alanine	aspartate	glycine	**U**
	valine	alanine	aspartate	glycine	**C**
	valine	alanine	glutamate	glycine	**A**
	valine	alanine	glutamate	glycine	**G**

Figure 2-7 Another type of RNA codon chart.

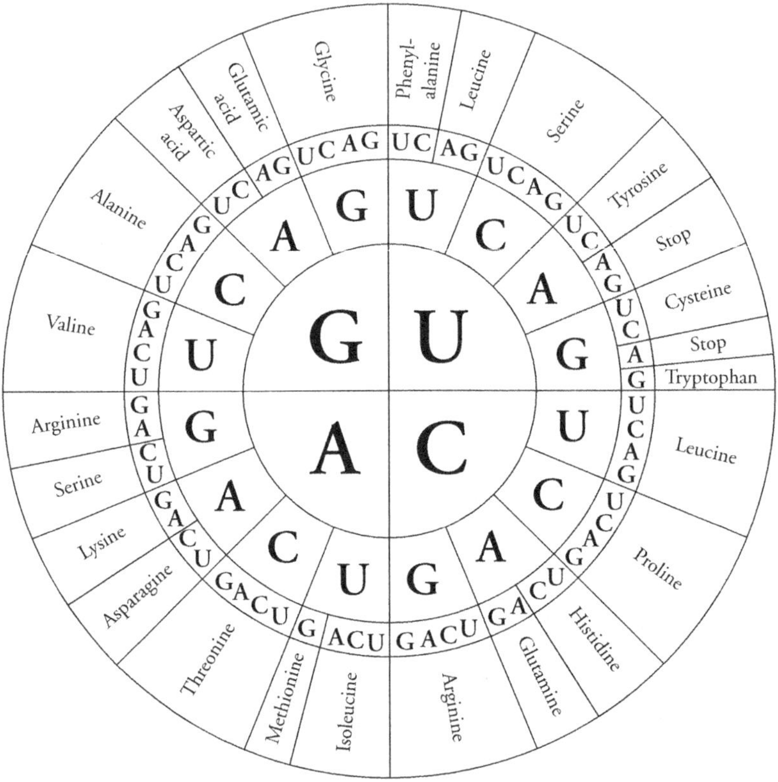

Let's look at how to read these two charts. In the first chart (Figure 2-6), begin by reading the first base, found on the left side; then read the second base, found at the top; and, finally, read the third base, found on the right side. So, if you have ACU, you would have the amino acid threonine. In the second chart (Figure 2-7), read it like a wagon wheel starting in the center. First, find the first base in the middle of the wheel; then, moving outward, find the second base in the middle circle; and, finally, find the third base on the outside circle. See if you can use this chart to find threonine from the code ACU.

Genetic mutations

Have you ever been to a restaurant, ordered from the menu, and had the wrong food brought to your table? A mistake was made somewhere between ordering your food and receiving your food. Possibly the waitress thought you said you wanted a steak when you actually wanted to order cake. Imagine your surprise when you are expecting dessert and a sirloin arrives at your table. Or perhaps the waitress heard you correctly but her handwriting was so terrible that the cook read it wrong and prepared the wrong food.

Mistakes like this can also happen during replication. Your body has many safeguards to prevent this, but if a mistake occurs and there is a permanent change to the organism's DNA, it is called a **genetic mutation.** Mutations can also occur due to mutagens. A **mutagen** is something in the environment that changes an organism's DNA. For example, too much exposure to ultraviolet (UV) light can cause changes to your DNA and can lead to skin cancer.

Some mutations can be permanent and can be passed down through generations from parents to offspring. DNA changes also can affect the protein being produced if the change in DNA then affects and changes the mRNA, which then gives the incorrect sequence and makes an incorrect protein.

Genetic Crosses

Mendelian genetics

In the mid-1800s, Austrian monk **Gregor Mendel** studied science at the University of Vienna and then worked in a monastery. One of his jobs was to maintain the monastery's garden. Mendel began doing thousands of experiments on the plants in the garden. Mendel is now known as the father of genetics. **Genetics** is the science of heredity or how traits are passed from parent to offspring.

One of the plants found in the garden was the garden pea. Garden peas grow and reproduce quickly, which made them a good scientific model for Mendel's studies. He was able to control the reproduction of the plant by removing the male part of the plant and fertilizing it with pollen from another plant of his choosing. One characteristic that Mendel studied was the height of the pea plant. He had some pea plants that produced short plants and other pea plants that produced tall plants. He also studied seed color, flower color, and seed shape.

One experiment he performed was to take a purebred plant that had yellow peas and fertilize it with the pollen of a purebred plant that had green peas. Purebred plants in nature will self-pollinate and produce offspring that are identical to their parents. Mendel controlled this by fertilizing the plants himself. Figure 2-8 shows an example of his experiment.

Figure 2-8 Mendel's pea plant experiment.

First generation
Green (yy)

Yellow (YY)	Yellow (Yy)	Yellow (Yy)
	Yellow (Yy)	Yellow (Yy)

↓

Second generation
Yellow (Yy)

Yellow (Yy)	Yellow (YY)	Yellow (Yy)
	Yellow (Yy)	Green (yy)

In genetics, we call the parent plant P, the first offspring F_1, the second offspring F_2, and so on. When the purebred plant that produced yellow peas was fertilized with the purebred plant that produced green peas, all of the F_1 generation had yellow peas. However, when Mendel next fertilized the F_1 generation with other F_1 plants, the results were a 3:1 mix of three yellow peas and one green pea. Mendel came to two conclusions based on his experiments. First, parents pass traits to their offspring and each offspring receives two genes—one from each parent. Second, some traits are dominant and some are recessive; this is known as the **principle of dominance.**

The things that are passed from the parent to the offspring are known as the organism's **genes,** and the different forms of the genes are called the **alleles.** So, the color of the pea plant would be the plant's genes, and if the plant had green or yellow peas, the color would be the plant's alleles.

Dominant/recessive

In Mendel's experiment with yellow and green peas, he discovered that yellow peas are dominant. You can think of a dominant trait as a trait that is stronger. Since yellow peas are dominant, when he crossed the purebred green pea plant with the purebred yellow pea plant, all the F_1 generation was yellow. This is because the offspring received one yellow pea plant gene from one parent and one green pea plant gene from the other parent. Therefore, since yellow was dominant, all the offspring were yellow. However, the offspring still had the green gene they had received from the green pea plant parent. This gene was just hidden and did not display itself in the plant's outward appearance.

However, when he then took the F_1 generation that were not purebred (they each had one green pea gene and one yellow pea gene) and crossed them, the results were different. In fact, for every three yellow pea plants, there was one green pea plant. Let's look at this in a different way.

The parent generation was purebred. Therefore, the yellow pea plant parents had two genes (one from each of their parents) and these genes were both yellow, while the green pea plants had two green genes. We will use a capital Y to stand for the dominant trait of yellow and a lowercase y to stand for the recessive trait of green. So, the yellow parent plant had two Y's (or YY) and the green parent plant had two y's (or yy). When crossed, the parents will give one gene to their offspring. The yellow plant will give a Y and the green plant will give a y. Because they are purebred, that is all they have to give.

The F_1 generation from this cross of YY and yy will all have the genes Yy, one from each parent. Since Y is dominant and stands for yellow pea plants, all the plants are yellow. However, they had a hidden green gene that was not seen. Next, when Mendel bred these now *not* purebred plants of Yy together, sometimes the plant would give a Y and other times it would give a y—almost like flipping a coin. It was just a matter of chance if the plant gave its Y or y gene. Look at Figure 2-9 to see this cross.

Figure 2-9 Punnett square: monohybrid cross.

Parent 2 genes ↓ / Parent 1 genes →	Y	y
Y	YY (Yellow)	Yy (Yellow)
y	Yy (Yellow)	yy (Green)

Punnett squares

A chart like the one in Figure 2-9 is called a **Punnett square.** Figure 2-9 is a simple Punnett square, looking at only a single gene, called a **monohybrid cross,** but there are others that can display even more information. To make a Punnett square, you simply take what is on the top of the column and what is on the side of the row and fill in the corresponding box. Doing so shows the possible combination of gametes from the two different parents. Each gamete (egg or sperm) contains only a single allele of each gene; when they combine through fertilization, you produce a potentially new genetic combination in the offspring. This is shown in each box.

Figure 2-10 is an example of a **dihybrid cross,** where two different genes are shown. The technique to fill out the chart is the same; it is just more complex because it shows more information.

Figure 2-10 Punnett square: dihybrid cross.

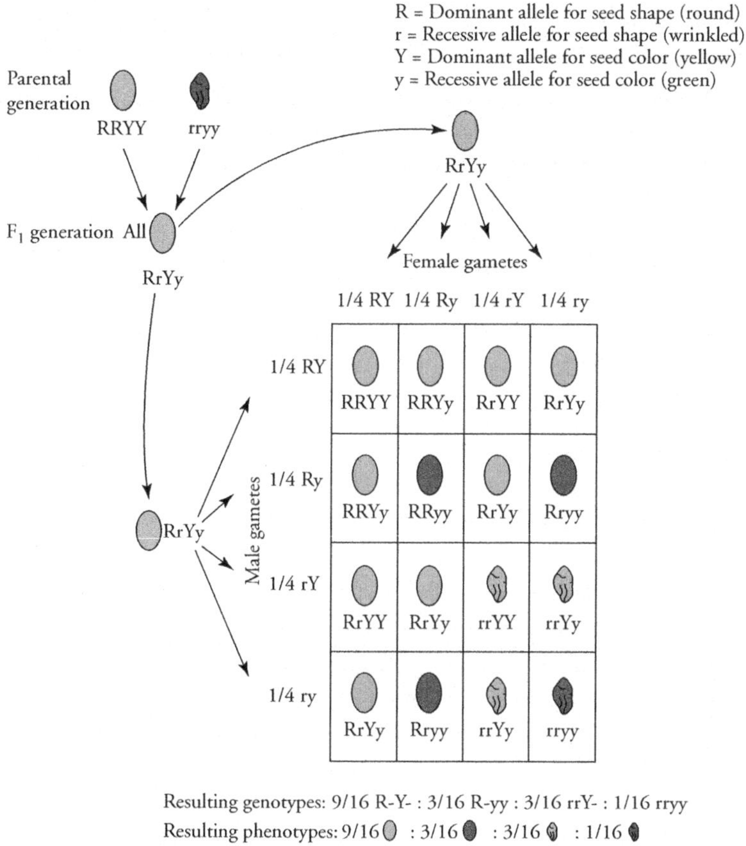

Genotype/phenotype

When we look at the traits of an organism and give it letters like Y for yellow and y for green, we are looking at the organism's genotype. The **genotype** of an organism is defined as the two alleles contained within that organism. The possible genotypes for pea color of the pea plant we discussed were YY, Yy, and yy.

The **phenotype** is the outward appearance of the genotype. So, YY is the genotype and yellow is the phenotype. In other words, the phenotype is what you see. You cannot look at a pea plant and tell that it is YY, but you can tell that it is yellow. Yellow is the phenotype in this example.

If an organism has a genotype of YY or yy, it is homologous for that trait. **Homologous** means that the alleles are the same. If the organism's genotype is Yy, it is **heterozygous,** meaning it contains two different alleles for the trait.

Non-Mendelian genetics

Mendel did very important work in helping us to understand our genes and how things are inherited; not all genes, however, are as simple as Mendel described. For example, if your father is 6 feet tall and your mother is 5 feet tall, most likely you are somewhere in the middle. Also, some organisms show variations of the genes, like color. Perhaps a white-flowered plant and a red-flowered plant are crossed; instead of getting all red or all white flowers, the offspring turn out with pink flowers. We refer to these types of examples as non-Mendelian genetics. Let's look at a few examples.

- **Polygenic inheritance** is when one gene is controlled by two or more sets of alleles (for example, eye color and skin color).
- **Intermediate/incomplete inheritance** is when neither original allele of a gene dominates alone. The alleles are "blended" to show a mix of the two. An example of this would be when a red flower is crossed with a white flower, and the resulting offspring is a pink flower.
- **Codominance** is similar to intermediate/incomplete dominance, but with codominance both alleles are equally dominant, not blended. An example of this is when a white flower is crossed with a red flower, and the offspring is red with white spots or white with red spots.
- **Multiple alleles** are genes with three or more different forms of the gene. Human blood is an example of a multiple allele. There are three different alleles for human blood type: A, B, and O. Each person will have only two of the three alleles—one from each parent.

- **Sex-linked inheritance** is the genetic link that deals mainly with the X and Y sex chromosomes that determine if you are male or female. However, there are other traits that are sex-linked, such as hemophilia, color blindness, fragile-X syndrome, and male patterned baldness. These are traits that are found on the X chromosome.

Meiosis

Each of your body cells contains 46 chromosomes (23 pairs). You received 23 chromosomes (one set) from your mother and 23 chromosomes (a second set) from your father. We refer to these as **homologous chromosomes,** one from each parent. During sexual reproduction, an offspring is produced that has a genetic mix from both parents. Just like you, your mother and father have 46 chromosomes, but if they both gave you 46, you would have 92, double the correct amount. So, there is a process that divides the 46 chromosomes from each parent in half so that each parent gives you only 23, resulting in your having the correct number: 46. This process is **meiosis.**

Diploid/haploid

When a cell has two copies of each chromosome, one from the mother and one from the father, it is known as **diploid.** When a diploid cell divides so that the resulting daughter cells have only one copy of each chromosome, that cell is known as **haploid. Meiosis** is the process of taking a diploid cell and dividing it into haploid cells, each haploid cell having only half the number of chromosomes. This ensures that after sexual reproduction, the offspring ends up with the correct number of chromosomes. Haploid cells are sperm produced by males and eggs produced by females.

Meiosis phases I and II

Cells go through two sets of division, known as meiosis I and meiosis II. The process of meiosis is similar to the process of mitosis that we learned about in chapter 1 (pages 22–24). Let's look at the steps of meiosis I first.

Meiosis I

Interphase I: Like mitosis, the first step of meiosis is interphase. During interphase I, the cells replicate their DNA, forming duplicate chromosomes, and prepare for division.

Prophase I: During prophase I, the nuclear membrane breaks down, and the replicated chromosomes condense and become visible. Chromosomes pair up with their homologous chromosomes in a process called **synapsis,** an event unique to meiosis. Spindle fibers begin to form, and the spindle fibers attach to the middle of the chromosome (called the centromere).

Metaphase I: During metaphase I, the homologous chromosomes line up in the middle of the cell. A difference between metaphase in mitosis and metaphase in meiosis I is that, in mitosis, single chromosomes line up, but in meiosis, the paired, homologous chromosomes line up (remember you received one from your mother and one from your father).

Anaphase I: During anaphase I, the spindle fibers pull the homologous chromosomes to the poles of the cell. Because, during metaphase I, your homologous chromosomes randomly lined up at the equator next to one another, your chromosomes mix up. In other words, in one cell you might have one chromosome from your mother on the right and one from your father on the left, and then in the next cell you might have one from your mother on the left and one from your father on the right. This makes it a random "shuffle" of your chromosomes that each new cell will have after meiosis I. The result will be that half the chromosomes went to one side and the other half went to the other side, reducing the number of chromosomes from $2n$ to n (half).

Telophase I: In telophase I, the spindle fibers begin to undo, the nuclear membrane forms again, and the cell performs cytokinesis (division of the cytoplasm), resulting in two cells, each of which has 23 chromosomes—some of the 23 from the mother and some from the father.

Figure 2-11 depicts meiosis I.

Figure 2-11 Meiosis I.

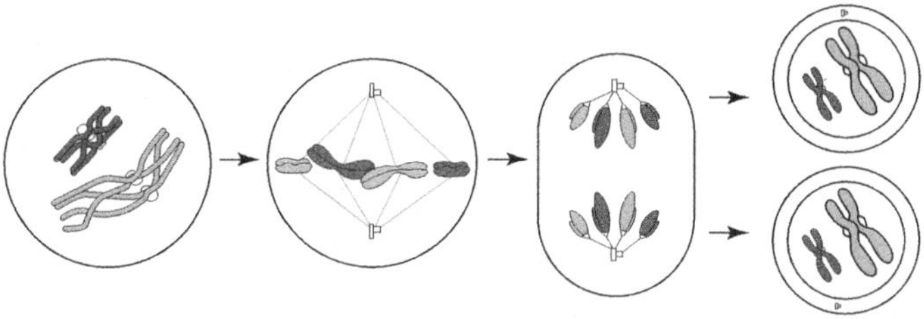

Meiosis II

Prophase II: During prophase II, the chromosomes condense, spindles form and attach to the chromosomes, and the nuclear membrane breaks down once again.

Metaphase II: During metaphase II, the chromosomes line up at the middle of the cell.

Anaphase II: During anaphase II, the sister chromatids separate and are pulled by the spindle fibers to the poles of the cell. Sister chromatids are the two halves of the replicated chromosomes.

Telophase II: During telophase II, the nuclear membranes form around each set of chromatids, the spindle fibers begin to undo, and cytokinesis (see page 24 in chapter 1) occurs.

Because meiosis II occurs in both of the cells that were formed after meiosis I, after meiosis II is finished, we now have four cells, all of which are haploid (*n*) and have some chromosomes from the mother and some from the father. Therefore, these cells are not identical, resulting in all the genetic differences we have in the world.

Figure 2-12 shows the steps of meiosis II.

Figure 2-12 Meiosis II.

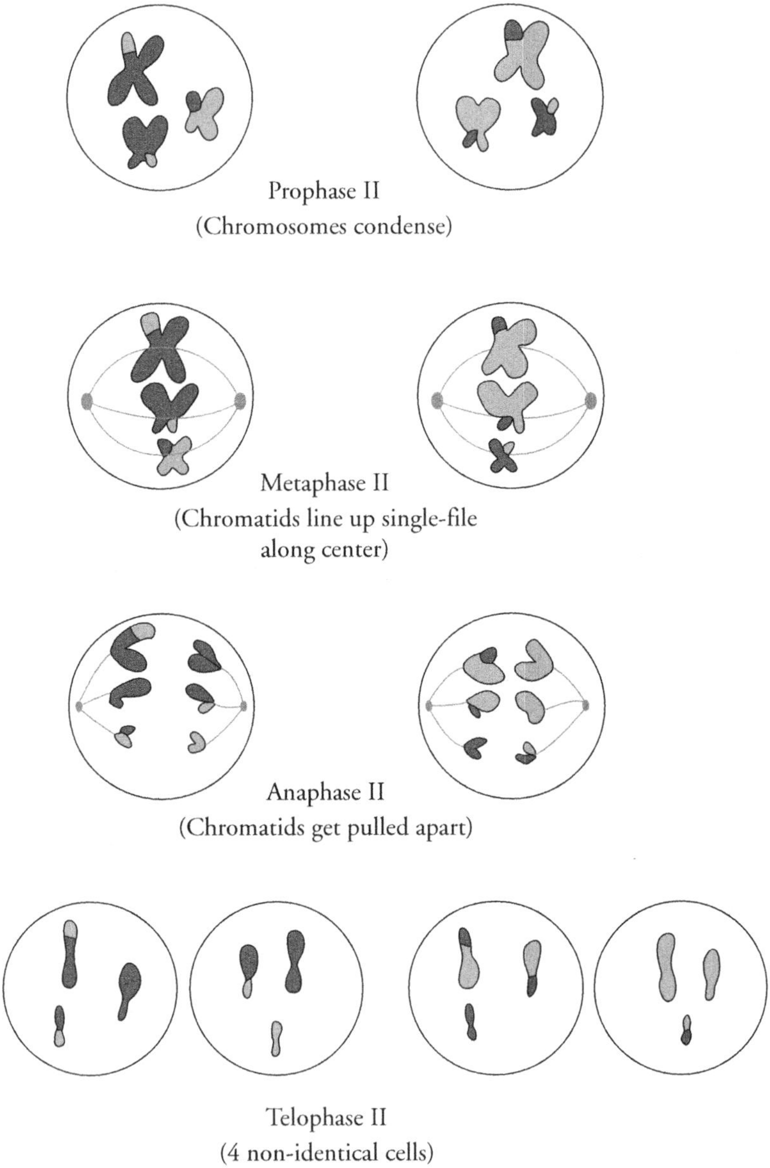

During meiosis, crossing over that results in an increase of genetic variation can occur. Crossing over is when parts of a chromosome exchange between a pair of homologous chromosomes. Figure 2-13 shows homologous chromosomes crossing over.

Figure 2-13 Homologous chromosomes cross over.

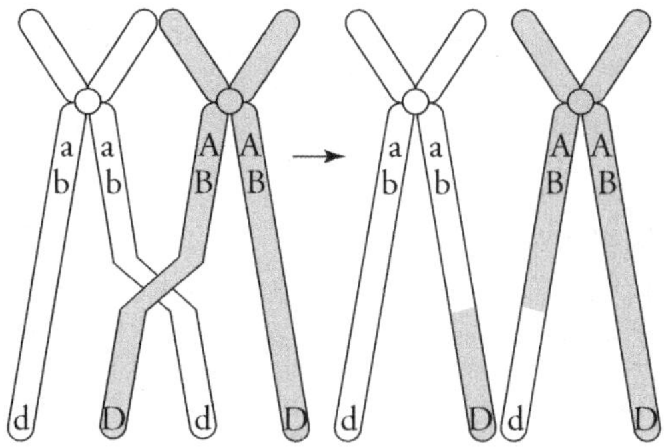

Let's look at a comparison chart of meiosis and mitosis (Figure 2-14).

Figure 2-14 Meiosis vs. mitosis.

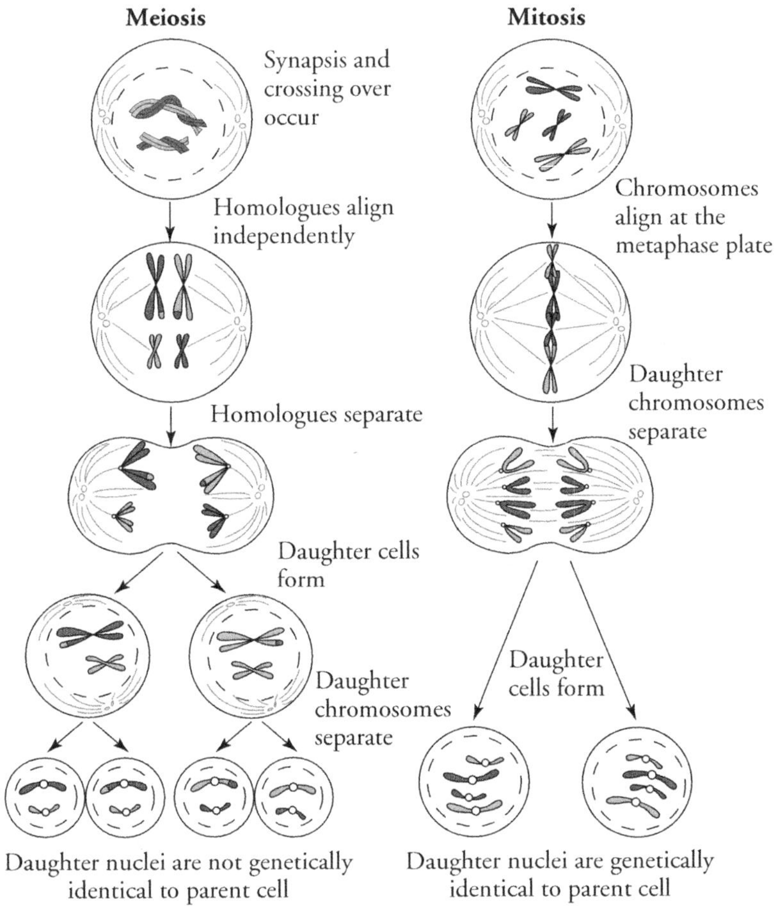

Karyotypes

A **karyotype** is a display of all the chromosomes of a cell. These are usually arranged by size and are stained with a special dye so that they can be photocopied through a microscope. Karyotypes are used to determine the sex of a fetus or to look for genetic disorders such as trisomy 21 (Down syndrome) or Klinefelter syndrome. Figure 2-15 is an example of a karyotype.

Figure 2-15 Karyotype.

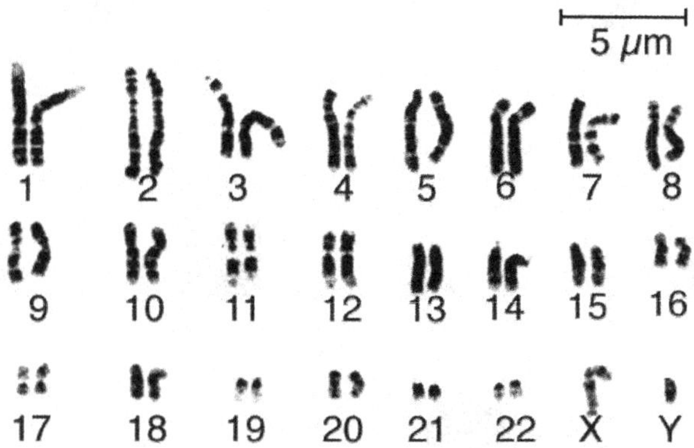

Pedigree

A **pedigree** is a display of the lineage or descent of an individual. This can help us see genetic traits and diseases across many generations in a family. Usually, females are drawn as a circle and males as a square, with lines connecting the parents and children. It is the same idea as a family tree. Figure 2-16 shows a pedigree.

Figure 2-16 Pedigree.

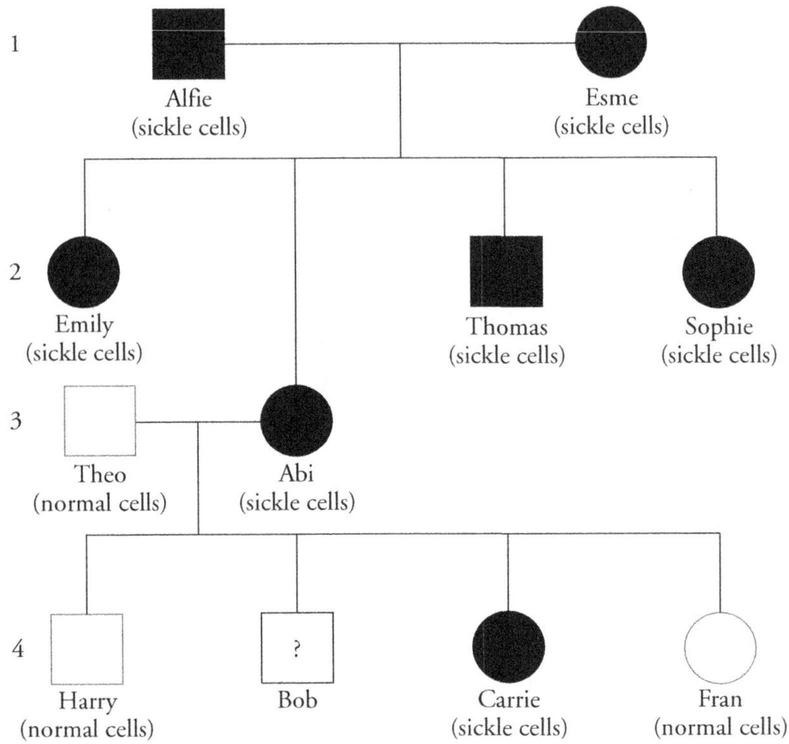

As you can see, oftentimes a pedigree is used to show disease and how the disease is transferred from parent to offspring.

Chapter Check-Out

1. A cellular process uses a strand of genetic material to produce a new strand. Parts of the strands are shown below.

   Original strand: CAG ATT
   New strand: GUC UAA

   This new strand will most likely be used for

   a. DNA synthesis.
   b. protein synthesis.
   c. crossing over.
   d. gene splicing.

2. Which amino acid would be translated from mRNA codon UUC?

First letter	Second letter				Third letter
	U	**C**	**A**	**G**	
U	phenylalanine	serine	tyrosine	cysteine	**U**
	phenylalanine	serine	tyrosine	cysteine	**C**
	leucine	serine	stop	stop	**A**
	leucine	serine	stop	tryptophan	**G**
C	leucine	proline	histidine	arginine	**U**
	leucine	proline	histidine	arginine	**C**
	leucine	proline	glutamine	arginine	**A**
	leucine	proline	glutamine	arginine	**G**
A	isoleucine	threonine	asparagine	serine	**U**
	isoleucine	threonine	asparagine	serine	**C**
	isoleucine	threonine	lysine	arginine	**A**
	(start) methionine	threonine	lysine	arginine	**G**
G	valine	alanine	aspartate	glycine	**U**
	valine	alanine	aspartate	glycine	**C**
	valine	alanine	glutamate	glycine	**A**
	valine	alanine	glutamate	glycine	**G**

   a. tryptophan
   b. asparagine
   c. arginine
   d. phenylalanine

3. A purebred animal with white fur is crossed with a purebred animal with red fur. The offspring exhibit both white and red fur. What type of inheritance pattern is displayed?
   a. incomplete dominance
   b. codominance
   c. sex-linked inheritance
   d. polygenic inheritance

4. The allele for blond hair is recessive (bb), and the allele for brown hair is dominant (BB). If a woman with blonde hair (bb) is crossed with a heterozygous man with brown hair (Bb), what is the probability that their child will have blond hair?
   a. 0 percent
   b. 25 percent
   c. 50 percent
   d. 75 percent

5. A section of a nucleic acid is shown below.

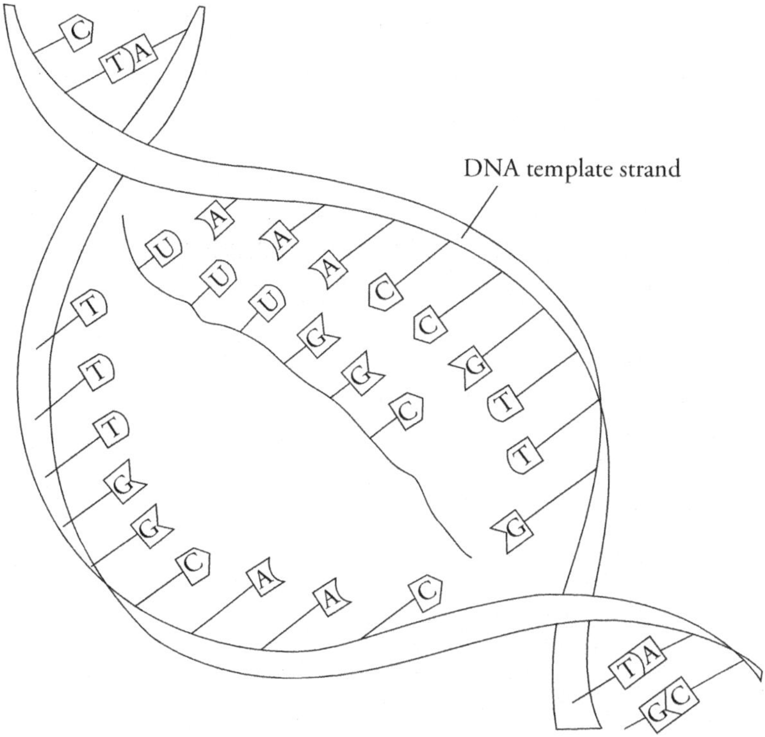

This picture shows a molecule being formed that is complementary to the strand of DNA. What molecule is being formed?

a. messenger RNA
b. duplicated DNA
c. protein
d. ribosome

6. The bonds between the complementary nucleotides in a double-stranded DNA molecule are what kind of bonds?

a. strong hydrogen bonds
b. strong ionic bonds
c. weak hydrogen bonds
d. weak ionic bonds

7. Genetic variability would most likely occur as a result of which of the following?
   - **a.** mitosis
   - **b.** protein synthesis
   - **c.** two homozygous individuals having a child
   - **d.** crossing over of homologous chromosomes during meiosis

8. Look at the chart below of the base sequences of various organisms.

Mitochondrial DNA Codes

Organism	***DNA Codes***
American black bear	ATT GGA GCA GAC TTA
Giant panda	ATT GGC ACT AAT CTA
Red panda	ATT GGA ACT AAC CTT
Raccoon	ATC GGA TCT AAC CTT

Which of the following organisms are the most closely related?
   - **a.** the raccoon and the American black bear
   - **b.** the American black bear and the giant panda
   - **c.** the giant panda and the red panda
   - **d.** the red panda and the raccoon

9. Look at the dihybrid cross below and predict the blank square.

	SY	Sy	sY	sy
SY	SSYY	SSYy	SsYY	SsYy
Sy	SSYy	SSyy	SsYy	Ssyy
sY	SsYY	SsYy	ssYY	ssYy
sy	SsYy	Ssyy	ssYy	

   - **a.** SSYY
   - **b.** ssYY
   - **c.** ssyy
   - **d.** SSyy

10. When DNA is copied, a base pair is sometimes substituted for a different base pair. Look at the diagram below and predict the outcome.

C	C		C	C
G	G		G	G
A	A		A	A
C	C		C	C
G	G		G	G
G	G		G	**A**
A	A		A	A
T	T	→	T	T
T	T		T	**G**
A	A		A	A
C	C		C	C
A	A		A	A
T	T		T	T

a. a mutation
b. a loss of the chromosome
c. a break in the chromosome
d. the synthesis of a protein

Answers: 1. b **2.** d **3.** b **4.** c **5.** a **6.** c **7.** d **8.** c **9.** c **10.** a

Chapter 3

BIOLOGICAL EVOLUTION AND CLASSIFICATION

Chapter Check-In

- ❑ Darwin's theory of evolution
- ❑ Evidence for evolution
- ❑ Evolutionary mechanisms
- ❑ Patterns of evolution
- ❑ Classification

Evolution is the idea that all life on Earth came from a common ancestor. Through time, because of many different processes acting on living things, we find the diversity of life we see on our planet today. Evolution occurs when there is a change in alleles in a population over time. These changes are then passed on to the next generation, and after many, many generations, we can see long-term change. One key thing to remember is that evolution takes multiple generations to occur; therefore, in the concept of time, evolution on Earth is very slow. In this chapter, we will look at the history of evolutionary theory, as well as mechanisms that effect change.

Darwin's Theory of Evolution

Darwin and history

Charles Darwin, born in England in 1809, proposed the concept we know today as evolution. **Evolution** is the theory of how living things change over time. In 1831, Darwin joined an expedition aboard the HMS *Beagle* that was mapping coastlines, particularly of South America. Darwin was a naturalist who loved to study the environment. He lived at a time when most people, including himself, believed that the world was about 6,000 years old and that living things had remained the same all that time.

The expedition lasted five years, and Darwin spent his time observing and collecting specimens of living things, as well as rocks and fossils. He began to pay attention to three things: species that are different around the world, species that are different in small areas, and species that change over time. When he returned from his voyage, he sent many of these specimens off to different scientists to be identified. The result of his observations and the evidence he collected after identification led Darwin to publish his book *On the Origin of Species by Means of Natural Selection* in 1859.

Around the same time, many other scientists were studying the same types of things around the world. The chart below summarizes the scientists and what they studied that helped to support Darwin's theory of evolution.

Scientist and Year of Importance	***Contribution***
Erasmus Darwin, grandfather of Charles Darwin (1794)	Suggested that all living things came from a common ancestor and that complex organisms came from less complex ones.
Carolus Linnaeus (1735)	Came up with a system of organization used to classify animals, plants, and minerals.
Georges Buffon (1749)	Examined organisms, variation among organisms, and evolution.
James Hutton (1785)	Studied geologic change and how mountains and valleys formed, concluding that Earth was much older than we thought.
Thomas Malthus (1789)	An economist who said that if the human population continued to grow at such a fast pace, we would run out of food and living space.
Jean-Baptiste Lamarck (1809)	Proposed the idea that organisms can change due to a desire to be perfect and become better over time. Though we now know this to be incorrect, Lamarck did show that species have not remained the same since time began.
Charles Lyell (1830)	Using information from Hutton and others, he published the book *Principles of Geology.*
Alfred Wallace (1858)	Studied where plants and animals are found and wrote a letter to Darwin that contained an essay about his theory of evolution. This theory was very similar to Darwin's natural selection theory.

Evidence for Evolution

As with all science, we must find evidence that can be supported for any theory to be considered. The evidence we use to support evolution is summarized in this section. A **theory** is something that has been tested over and over and is well supported by evidence from nature.

Fossil record

When Darwin was in Argentina, he discovered a fossil of a giant armadillo, known as a *Glyptodon.* He noticed that the fossil armadillo and the modern-day armadillo had many characteristics that were very similar (see Figure 3-1). This led Darwin to question if the modern armadillo had descended from the ancient armadillo but had changed over time. It also made him suspect that Earth was older than the assumed 6,000 years because the changes he was proposing would likely take much longer than 6,000 years to occur.

Figure 3-1 The picture on the left is an artist's rendition of a *Glyptodon,* and the picture on the right is a modern-day armadillo.

Darwin had many experiences like this one, where he found fossils in places where they seemed unlikely to occur (like marine organisms at the top of very tall mountains). He also experienced an earthquake that lifted the shore 3 meters up and out of the sea. He reasoned again that changes like this could support his theory.

Anatomy: Homologous, vestigial, and analogous structures

One of the key pieces of evidence that Darwin used is known as homologous structures. **Homologous structures** are parts of an organism's anatomy that have a similar structure and come from a common ancestor. It is important to understand that although their structures are similar, the structures might not have similar functions. Look at the examples in Figure 3-2 below.

Figure 3-2 Homologous structures in various animals, showing similar structure but not similar function.

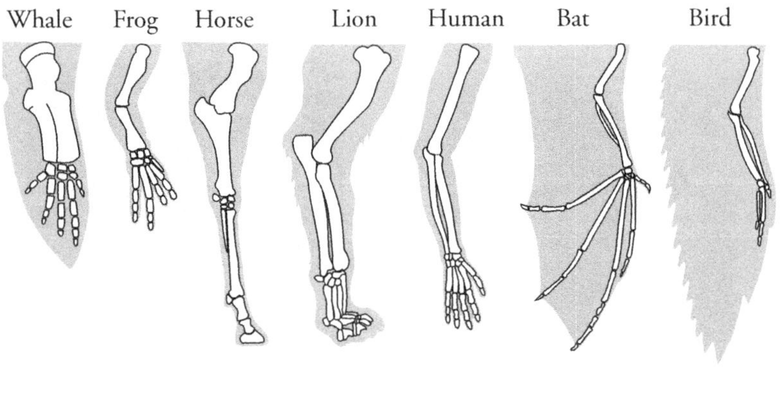

In Figure 3-2, you can see that the bones in the animals all have similar structures but yet do very different things. For example, a horse's leg and a bat's wing do not perform the same functions, but the bones look similar in structure.

Another piece of evidence Darwin used is vestigial structures. A **vestigial structure** is something that was passed down from an ancestor but has lost most of the size and function of the structure. For example, in humans the appendix is thought to be a vestigial structure. We still have a very small appendix (see Figure 3-3) that does very little for us (except cause us problems if it becomes infected or ruptures) and is assumed to have been passed down from a common ancestor that ate large amounts of cellulose from plants. Since humans do not consume large amounts of cellulose today, the appendix has shrunk and remains relatively useless to us.

Figure 3-3 The appendix is a vestigial structure that does very little for us but was passed down from a common ancestor.

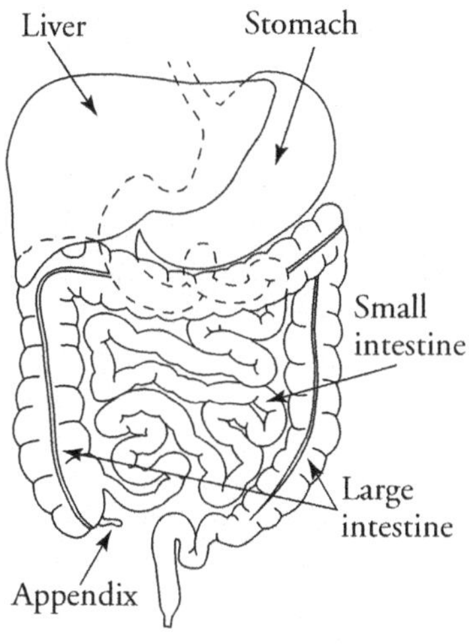

An analogous structure is another clue to evolution. **Analogous structures** perform the same function but have different structures. For example, think of a bird's wing and a butterfly's wing. A bird's wings have bones with skin and feathers around them that help the bird to fly. Similarly, a butterfly can fly, but its wings do not have bones, skin, or feathers. So, the wings of both organisms help the organism to fly, but they are very different structures (see Figure 3-4). Analogous structures are a clue to evolution because the species live in similar environments and fill similar ecological roles. The species are not related to one another. Their similar traits show that they evolved in different species, instead of from a common ancestor, because it was the best adaptation for the environment they inhabited.

Figure 3-4 Analogous structures, like a butterfly wing and a bird wing, have similar functions but different structures.

Biogeography

Now that we have looked at the anatomy of different organisms that help to support Darwin's theory of evolution, let's look at another piece of evidence known as biogeography. **Biogeography** is the study of where living things are found today and where they were found in the past (by looking at the fossil record).

Darwin studied finches on the Galapagos Islands. He noticed that, depending on where the finches were found, they had different characteristics. For example, some finches had short, pointy beaks for eating plants, while others had longer, stronger beaks for breaking large, hard seeds. However, the birds all had a common ancestor from the mainland of South America. He began to wonder if these birds had migrated from the mainland and then, with different environmental conditions, different food choices, and different behaviors (depending on where on the islands they lived), had evolved differently to better survive in their particular environments. This information helps us to understand how today's organisms evolved from their ancestors.

Molecular similarities

There is evidence for evolution at the molecular level as well. For example, when we look at the DNA, RNA, and proteins of organisms, we find the same biochemical properties—whether we are studying a rabbit or a bullfrog or even a yeast. The fact that these molecular compounds are similar in all living things leads to evidence of a common ancestor.

Developmental similarities

When we look at the embryos of vertebrates, we can see that they look very similar. For example, a human and a fish both have gill slits, but the gill slits turn into gills in fish and disappear in humans before birth. When looking at animal embryos, it is often very hard to tell them apart or tell what they will turn into. This developmental evidence is another piece of the evolution puzzle because common developmental patterns reflect evolutionary kinship. The similarities last longer the more closely related the two organisms are. See Figure 3-5 for different animals and how they look in early embryonic development and how they change as they develop. Also note how organisms that are closely related (like the mammals) look similar longer and how organisms that are less closely related (like mammals and fish) begin to look different as they mature.

Figure 3-5 How different animals look during early embryonic development. As you can see, animals that turn out looking very different as adults look very similar early on in their development.

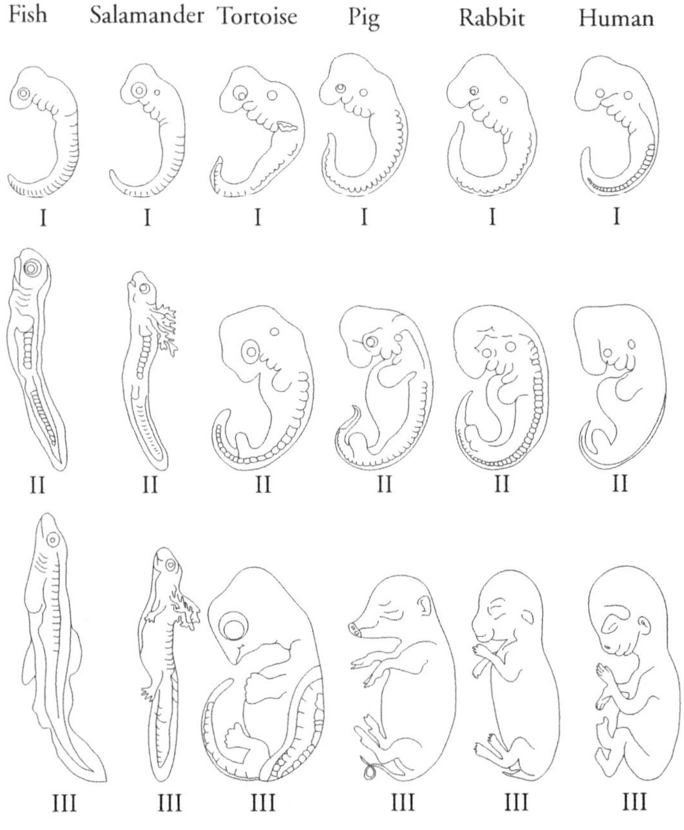

Evolutionary Mechanisms

Now that we have looked at the evidence and the history of evolution, let's look at the mechanisms that drive the evolutionary processes. The five mechanisms we will study are genetic drift, gene flow, mutations, sexual selection, and natural selection. We'll also study speciation.

Genetic drift: Bottleneck and founder effects

Genetic drift is the change in allele frequency that is random and can lead to a loss of genetic diversity in a population. It is caused by chance and is selected through random, independent assortment. There are two types of genetic drift: the bottleneck effect and the founder effect.

The **bottleneck effect** occurs when a natural disaster wipes out a large percentage of the population, leaving only a few individuals to reproduce. For example, imagine there is a population of beetles living in a forest and a forest fire comes through, killing 95 percent of the beetles and leaving only the remaining 5 percent to reproduce. If these 5 percent were genetically different from the majority of the original population, the surviving population would be very different from the original. See Figure 3-6 below as an example.

Figure 3-6 The bottleneck effect. In the parent population, we have an equal number of gray and white balls. However, most of the surviving balls after a "disaster" were gray, so the next generation has a much greater number of gray than white.

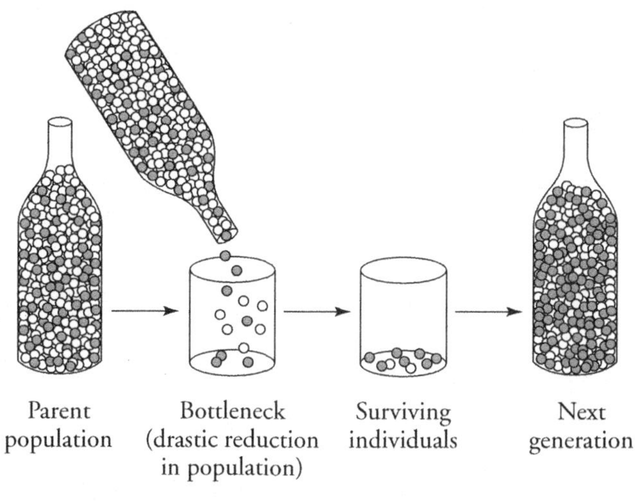

The other method of genetic drift is the **founder effect,** where a few individuals from an original population migrate to and inhabit a new environment (see Figure 3-7). These few individuals will reproduce without the large gene pool from the original population, which could change the allele frequency of the population. Again, if, say, 5 percent of organisms were different from the original population, the new population would be very different. The best example of this is on islands, when only a few individuals from the mainland make their way to the island and establish a new colony.

Figure 3-7 The original population of beetles had a pretty equal number of gray- and white-colored individuals. However, the founders (the individuals that went to a new environment) were all gray beetles, so the new generation on the island consisted of mostly gray beetles.

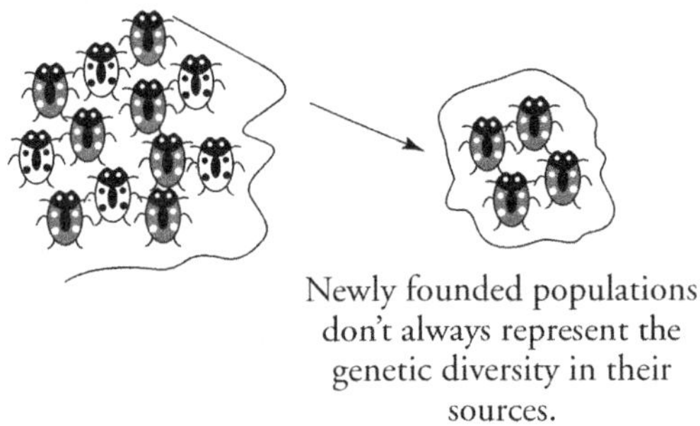

Gene flow

Gene flow is the transfer of alleles from one population to another. Migration into or out of a population may be responsible for the change in allele frequencies. This random movement can cause evolution to occur. See Figure 3-8.

Figure 3-8 Two populations are migrating into and out of the two areas, exchanging allele frequencies.

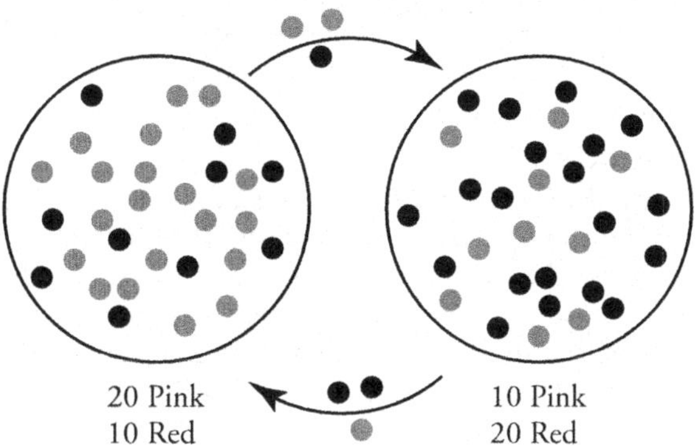

Mutations

Recall from chapter 1 that **mutations** are changes in a cell's DNA. These changes are usually harmful, but can sometimes give the organism an advantage. If this advantage is then passed on from one generation to the next, it can become more common and cause evolutionary change. An example of a positive advantage is the sickle-cell trait that protects a person against malaria. The sickle-cell trait warps the shape of a normal round red blood cell into one that is more crescent moon–shaped. Scientists do not fully understand why, but they think that it could be because the shape of the person's blood cells does not allow the malaria parasite to infect the cell. It seems that people who carry the sickle-cell trait have been naturally selected because of the protection they get from malaria.

Sexual selection

In some species, the ability to attract a mate, known as **sexual selection,** can lead to evolutionary change. This is often common in birds, where the male bird is brightly colored and the female is plain. The male bird will often do a display of his feathers or a mating ritual to attract a female to mate. If the male with the brighter, longer, fancier feathers has a greater chance of reproducing, then over time, the offspring will have these traits, and evolution occurs.

Natural selection

The final mechanism for evolution we will discuss is natural selection. **Natural selection** is when an organism has traits that allow it to survive and reproduce in nature, so these traits are passed from one generation to the next. For example, if there is a beetle species in which some beetles are green and others are brown, and the brown beetles are more likely to blend into the leaves and the trees, while the green ones stand out, the green ones are more likely to be found and eaten by predators (see Figure 3-9). These green beetles are now no longer able to reproduce because they were killed, while the brown ones continued to reproduce. After many generations, the population will have more brown beetles than green beetles.

Figure 3-9 Natural selection is shown where the white individuals were more likely to survive in their environment and, therefore, were able to reproduce and pass their genes onto the next generation.

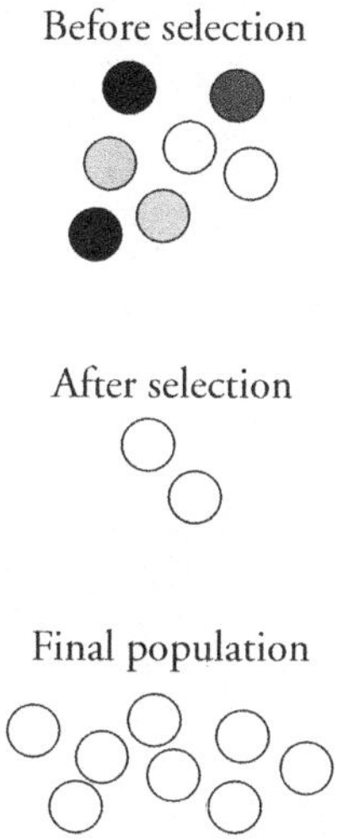

Speciation

The definition of a **species** is the largest group of organisms capable of interbreeding and producing fertile offspring. **Speciation** is the evolutionary process by which new biological species form. If members of the species stop breeding with one another, we say they have become **reproductively isolated** from one another. Reproductive isolation can occur by behavioral, geographic, and temporal isolation.

Behavioral isolation

Behavioral isolation occurs when the reproductive behaviors of two species no longer allow for mating. For example, if two bird populations have different songs to call each other and to signal a mating ritual and the two songs no longer attract the two populations, then the populations will not mate, which can eventually lead to two separate species.

Geographic isolation

Geographic barriers can lead to reproductive isolation. If two populations become separated by a river or a mountain, they may become **geographically isolated** from one another and no longer be able to physically meet to reproduce (see Figure 3-10). Over time, these populations may become two separate species.

Figure 3-10 The original beetle population became isolated by a river that separated the population into two. The two new populations were unable to reproduce with one another and, over time, became two different species of beetles.

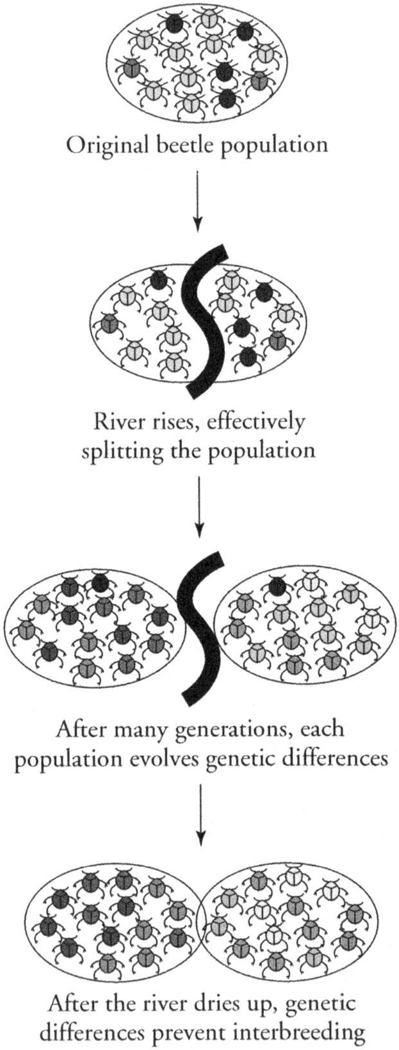

Temporal isolation

Temporal isolation is when two populations no longer reproduce at the same time. For example, if members of one population are ready to mate in the spring and the members of the other population are not ready to mate until the fall, then the two populations may stop breeding and become two different species.

Patterns of Evolution

Looking at the patterns that may have led to change is another way scientists try to find evidence for evolution. We will now look at some of these evolution patterns.

Divergent and convergent evolution

Divergent evolution (sometimes called adaptive radiation) is when one species diversifies into many different species due to different environmental pressures and conditions. For example, evidence of a common ancestor exists in the kit fox and the red fox. The kit fox is sandy-colored and blends in well to desert habitats, while the red fox has coloring that helps it to blend into forested areas. However, both fox species have anatomical structures similar enough to be evidence of evolving from a common ancestor.

Convergent evolution is when two species that live in similar environments (for example a desert in Africa and a desert in the United States) evolve to have similar structures to help them survive, even though they are completely different species. An example would be flying insects, bats, and birds that have all adapted to be able to fly. Another example would be anteaters from Australia, North America, and Africa. These anteaters are not closely related genetically, but they have evolved traits that allow them to eat ants—a long, sticky tongue, few teeth, a tough stomach, and large salivary glands. These developments occurred separately but allow them to eat the same kinds of organisms.

Gradualism and punctuated equilibrium

Gradualism is the pattern of evolution where species evolve slowly over long periods of time. This is different from another pattern of evolution called **punctuated equilibrium,** where species remain relatively stable for long periods of time and then go through quick bursts of evolutionary change. It is believed that punctuated equilibrium would occur when a disaster (such as a volcanic eruption) to an ecosystem happens, resulting in species evolving quickly to adapt to the new environment. See Figure 3-11.

Figure 3-11 On the left is a picture of gradualism, where one common ancestor evolved slowly over time. The picture on the right is a representation of punctuated equilibrium, where one common ancestor went through quick bursts of change, probably due to a disaster in the ecosystem.

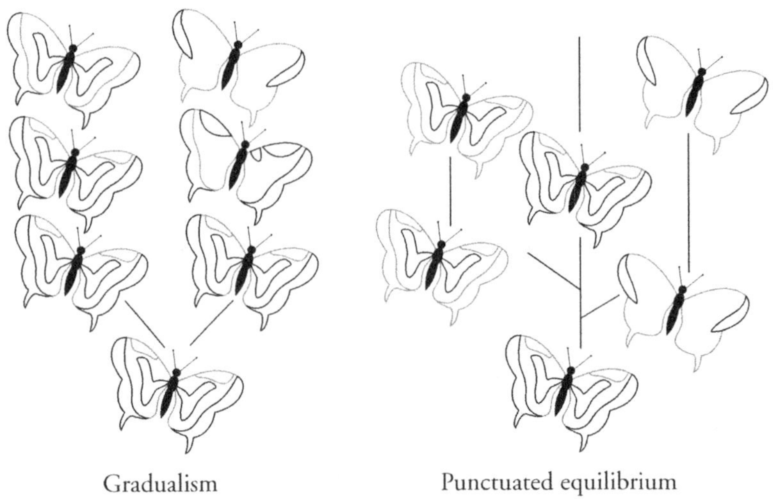

Genetic variation

There are a couple of ways that genetic variation can occur. The first is a chance mutation. A **mutation** is when there is a change in the DNA of an organism. As mentioned earlier, most of the time, this is a negative change and does not result in evolutionary change, but sometimes an advantageous change occurs that is then passed down to future generations and can effect evolutionary change. The second way that genetic variation can occur is through **recombination,** the reshuffling of genes during meiosis that results in different genetic combinations and can lead to long-term change.

Classification

We classify things so that we can put them into groups based on how they are alike. Science uses a system of classification called taxonomy to group living things. **Taxonomy** is the scientific way of putting organisms with similar characteristics into groups. For example, all mammals are put into the same group known as a class, and all vertebrates are put into a group known as a phylum.

Taxonomy

Carolus Linnaeus was a botanist who came up with a way to classify living things so that scientists everywhere would know what specimen they were studying. He developed a two-word naming system known as **binomial nomenclature** that gives each species a scientific name. The first word represents the genus, and the second word represents the species. He also used Latin as the language because it does not change and is used in science. The **genus** (the first word) of the organism is capitalized, and the **species** (the second word) is lowercase; both words are italicized. So, for example, a human is *Homo sapiens. Homo* is the genus, and *sapiens* is the species.

Hierarchical classification system

Over time, the two-name system expanded to include seven groups known as **taxa.** The largest group is the **kingdom.** There are six kingdoms: Plantae, Animalia, Protista, Fungi, Archaebacteria, and Eubacteria. Any multicellular heterotroph whose cell structure does not include a cell wall would be in the kingdom Animalia.

Just think of all the different animals on the planet. We need to be able to further narrow down an organism so we can continue to identify it. The next level is the phylum. An example of **phylum** is Chordata; any animal with a vertebrate, like a human, would fall into this phylum. The next, less-inclusive level is **class.** An example of class would be Mammalia, or any animal that is a mammal. We are now getting to the level of **order.** Humans are in the order primates. **Family** is next. Humans are members of the hominidae family. Last, we get to the genus and species levels. As you can see, the levels become less inclusive as you move from kingdom to genus. Think about mammals. There are many different mammals out there, from humans to dogs to polar bears. Humans have similarities with dogs and polar bears, but humans are not the same species. In fact, we are not the same genus, family, or order; our similarities end at the class Mammalia. See Figure 3-12 for the taxonomy of a human.

Figure 3-12 Human taxonomy. The chart below shows the taxonomic categories and the names for each level for a human.

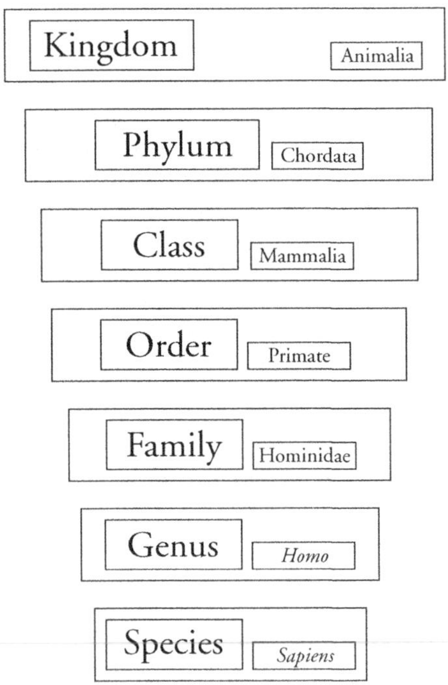

Cladograms

A **cladogram** is a way to show evolutionary relationships among groups of organisms. Think back to earlier in this chapter when we talked about the different ways organisms evolved (divergent and convergent evolution; gradualism and punctuated equilibrium). A cladogram shows where the organisms split from a common ancestor. See Figure 3-13.

Figure 3-13 This cladogram shows the evolution from a common ancestor to a wide range of organisms living today.

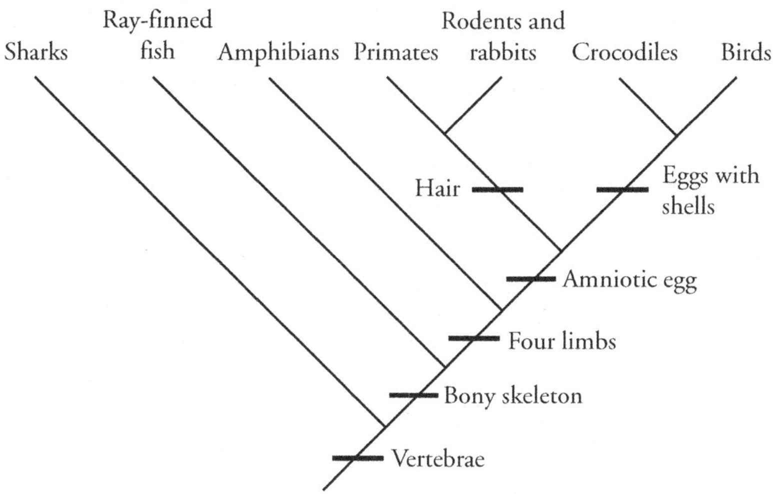

As you can see in the cladogram, a common ancestor with a vertebrae gave rise to many different organisms living today. Each time an organism branches off is the last time the organisms shared a common ancestor. Scientists can use cladograms to show the number of characteristics shared by a group and how recently the group shared a common ancestor.

Dichotomous key

A **dichotomous key** lets a scientist determine the identity of new things in the natural world. *Di* means "two," so a dichotomous key always gives two choices in each step. Look at the dichotomous key in Figure 3-14 and follow the directions for using it.

1. Start at Step 1. You have the leaf you are trying to identify at the left. Is the leaf compound or simple? It is a simple leaf, so we go on to Step 4.
2. Step 4 asks about the arrangement of leaf veins. Since our leaf has veins that branch off the main vein in the middle of the leaf, we go on to Step 6.
3. Step 6 asks about the appearance of the leaf edge. The leaf has a jagged edge, so our leaf is a *Betula* (birch).

Figure 3-14 Dichotomous key.

Dichotomous Key for Leaves

1. Compound or simple leaf
   1a) Compound leaf (leaf divided into leaflets) .... go to step 2
   1b) Simple leaf (leaf not divided into leaflets) .... go to step 4
2. Arrangement of leaflets
   2a) Palmate arrangement of leaflets (leaflets all attached at one central point) ....*Aesculus* (buckeye)
   2b) Pinnate arrangement of leaflets (leaflets attached at several points) .... go to step 3
3. Leaflet shape
   3a) Leaflets taper to pointed tips ....*Carya* (pecan)
   3b) Oval leaflets with rounded tips ....*Robinia* (locust)
4. Arrangement of leaf veins
   4a) Veins branch out from one central point .... go to step 5
   4b) Veins branch off main vein in the middle of the leaf .... go to step 6
5. Overall shape of leaf
   5a) Leaf is heart-shaped ....*Cercis* (redbud)
   5b) Leaf is star-shaped ....*Liquidambar* (sweet gum)
6. Appearance of leaf edge
   6a) Leaf has toothed (jagged) edge ....*Betula* (birch)
   6b) Leaf has untoothed (smooth) edge ....*Magnolia* (magnolia)

Chapter Check-Out

1. The limbs of different organisms are shown below. How are these limbs evidence of a common ancestor?

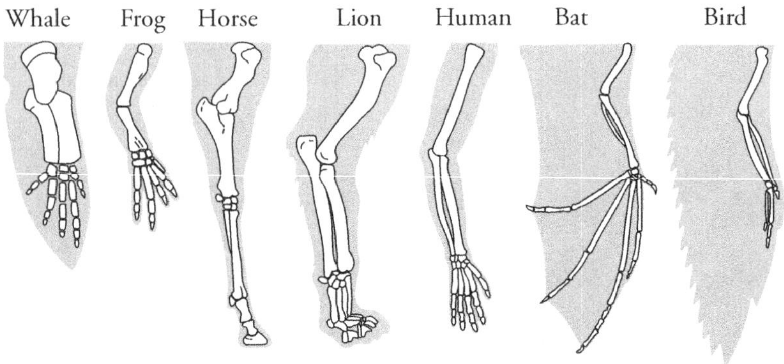

   a. They all perform the same basic function.
   b. They all have the same basic structure.
   c. They are all mammals.
   d. They all are used for an animal's movement.

2. There are anteaters found in Africa, North America, and Australia. These animals all have adaptations to allow them to eat and digest ants.

   The fact that these animals have similar dietary characteristics is evidence that

   a. they have evolved traits that allow them to eat ants.
   b. they are all mammals.
   c. a mutation in their DNA allowed for these traits to form.
   d. a bottleneck effect occurred that allowed these characteristics to form in anteaters.

3. The following cladogram shows the evolution of different animal species from the fossil record.

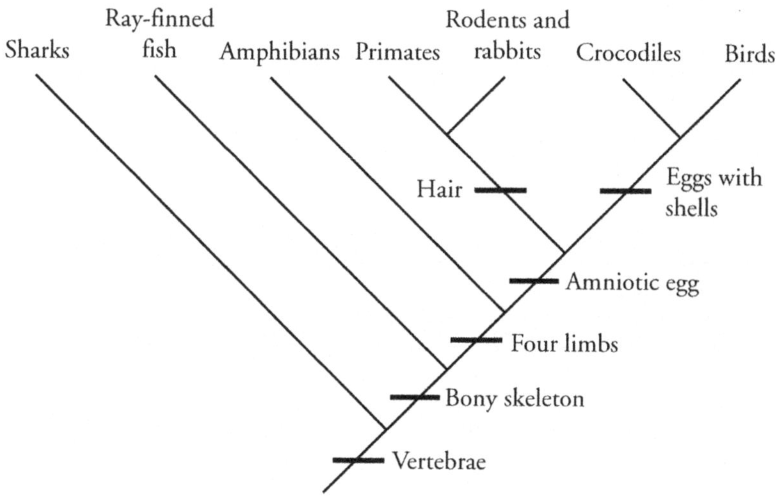

   Which of the following scenarios would change the legitimacy of this cladogram?

   a. a fossil of a primate with a bony skeleton
   b. a fossil of a rodent with a vertebrae
   c. a fossil of a shark with an amniotic egg
   d. a fossil of a crocodile with eggs with shells

4. Using a standard hierarchal system allows scientists from all over the world to

a. determine which common ancestor an organism came from.
b. understand the ecological pyramid of the different organisms.
c. use a common way to classify organisms.
d. understand if they have similar skeletal structures.

5. Carolus Linnaeus came up with a system to classify organisms. Which of the following levels shows the closest relationship?

a. kingdom
b. order
c. class
d. phylum

6. Use the dichotomous key below to identify Bird Z.

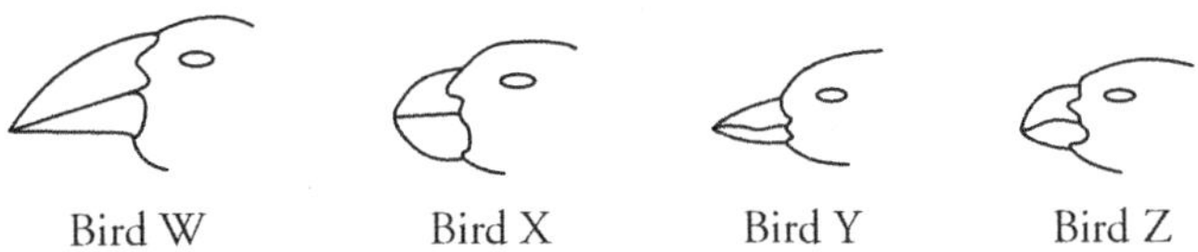

Dichotomous Key to Representative Birds	
1. a. The beak is relatively long and slender	*Certhidea*
b. The beak is relatively stout and heavy	go to 2
2. a. The bottom surface of the lower beak is flat and straight	*Geospiza*
b. The bottom surface of the lower beak is curved	go to 3
3. a. The lower edge of the upper beak has a distinct bend	*Camarhynchus*
b. The lower edge of the upper beak is mostly flat	*Platyspiza*

According to the key, Bird Z is what genus of bird?

a. *Certhidea*
b. *Geospiza*
c. *Camarhynchus*
d. *Platyspiza*

7. If a mutation in an organism's genes helps it to survive and reproduce, what will probably occur in future generations?

a. The mutation will spread to other species.
b. The mutation will increase the individual's lifespan.
c. The mutation will increase the number of genes in the species.
d. The mutation will show up more in the genes of the population.

8. The table below shows the classification of four different butterflies.

	Organism 1	***Organism 2***	***Organism 3***	***Organism 4***
Common Name	Orange-barred sulphur	Orange-banded protea	Silver-spotted flambeau	Silver-studded blue
Class	Insecta	Insecta	Insecta	Insecta
Order	Lepidoptera	Lepidoptera	Lepidoptera	Lepidoptera
Family	Peridae	Lycaenidae	Nymphalidae	Lycaenidae
Genus	*Phoebis*	*Capys*	*Agraulis*	*Plebejus*

According to the table, what butterflies are the most closely related?

a. 1 and 2
b. 2 and 3
c. 2 and 4
d. 1 and 3

9. When the fossil record remains relatively unchanged for long periods of time and then we have quick, short periods of significant evolutionary change, we are witnessing what kind of evolution?

a. punctuated equilibrium
b. gradualism
c. genetic drift
d. the founder effect

10. A farmer sprays his crops with a pesticide, and 90 percent of the bugs are killed. The next year, he sprays with the same pesticide, and only 50 percent of the bugs are killed. What most likely happened that caused this pesticide to not work as well the second year?
    a. New bugs moved into the farmer's field that the pesticide could not kill.
    b. A few of the bugs survived the first year and reproduced, so the new generation of bugs carried the allele for resistance to the pesticide.
    c. The pesticide stopped working because of a change in the chemical formula by the company.
    d. The farmer did not properly spray his crops the second year.

Answers: 1. b **2.** a **3.** c **4.** c **5.** b **6.** c **7.** d **8.** c **9.** a **10.** b

Chapter 4

BIOLOGICAL PROCESSES AND SYSTEMS

Chapter Check-In

- ❑ Photosynthesis and cellular respiration
- ❑ Animal systems
- ❑ Plant systems

Plants and animals have physiological systems that help them maintain homeostasis, the tendency of a biological system to remain internally stable. These systems interact to help keep the organism alive and functioning. In this chapter, we will study these systems in detail.

Photosynthesis and Cellular Respiration

Photosynthesis

Reactants and products

Photosynthesis is a process used by plants, as well as some bacteria and protists, to convert the energy of the sun into a sugar called glucose. These sugars can then be stored for later use or used immediately for energy or growth.

The reactants of photosynthesis (water and carbon dioxide) are the products of cellular respiration; the reactants of cellular respiration (glucose and oxygen) are the products of photosynthesis. Figure 4-1 shows how both of these metabolic processes are linked together. Keep in mind that **reactants** are substances initially present in a chemical reaction that are consumed during the reaction to make products. The **products** are compounds that are formed when a chemical reaction goes to completion. Think of when you make chocolate chip cookies. The ingredients like the eggs, flour, chocolate chips, and butter are the reactants, and the finished chocolate chip cookies are the products. In photosynthesis, carbon dioxide, water, and energy from the sun are used to make glucose and oxygen. In cellular respiration, the glucose and oxygen are used and carbon dioxide, water, and energy are released.

Figure 4-1 How the reactants of photosynthesis (the left side of the arrow) become the products (the right side of the arrow) of cellular respiration and vice versa.

Photosynthesis

$$\underset{\text{Carbon dioxide}}{6CO_2} + \underset{\text{Water}}{6H_2O} + \underset{\text{Energy in}}{\text{Energy}} \longrightarrow \underset{\text{Glucose}}{\boxed{C_6H_{12}O_6}} + \underset{\text{Oxygen}}{6O_2}$$

Cellular Respiration

$$\underset{\text{Glucose}}{\boxed{C_6H_{12}O_6}} + \underset{\text{Oxygen}}{6O_2} \longrightarrow \underset{\text{Carbon dioxide}}{6CO_2} + \underset{\text{Water}}{6H_2O} + \underset{\text{Energy out}}{\text{Energy}}$$

Light reactions

Let's look at photosynthesis in greater detail now. Keep in mind that photosynthesis is the way that light energy from the sun is converted to chemical energy. The formula for photosynthesis is found in Figure 4-2.

Figure 4-2 The formula for photosynthesis.

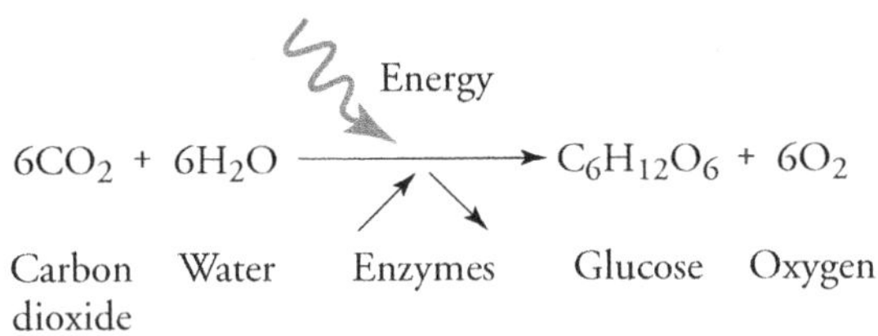

We will focus on photosynthesis in plants during this discussion rather than photosynthesis in bacteria or protists. In a plant, there are membrane-bound organelles called chloroplasts where photosynthesis occurs. These are mainly found in the leaves of plants, but they can be found elsewhere in some cases. There are two phases of photosynthesis: the light-dependent reactions and the light-independent reactions. Let's look at the light-dependent reactions first.

The **light-dependent reactions** use energy from the sun to produce energy-rich compounds like adenosine triphosphate (ATP). ATP is a molecule that supplies energy for the creation and arrangement of new covalent bonds. When six carbon dioxide molecules are used to build a single molecule of glucose, there are many new bonds being formed, and this requires a lot of energy. All of this occurs in the chloroplasts, but specifically on structures inside the chloroplasts known as thylakoids. Thylakoids look like stacks of coins, as you can see in Figure 4-3. Stacks of thylakoids are known as **grana.**

Figure 4-3 A chloroplast. As you can see, there are coin-shaped objects known as thylakoids inside the chloroplast. Energy is transferred along the thylakoid membrane.

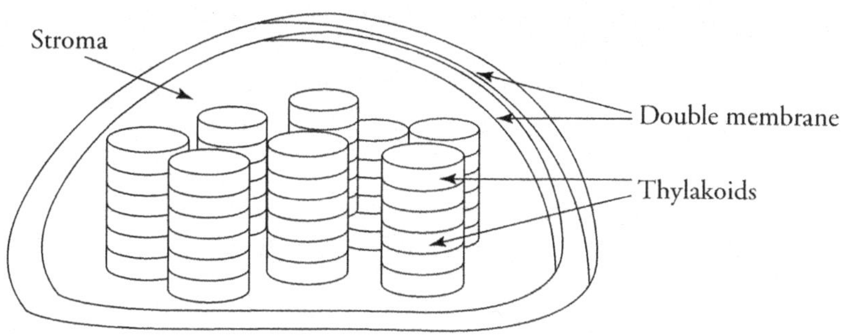

The **light-independent reactions** (also known as the **Calvin cycle**) use energy from the light-dependent reactions to make sugars. The plant takes carbon dioxide from the atmosphere and, using the ATP that was produced during the light-dependent reactions, produces the sugar glucose. These reactions take place in the stroma. The **stroma** is the fluid-filled space inside the chloroplast. Again, refer to Figure 4-3 to locate the stroma. The light-independent reactions do not need sunlight and can occur at any time as long as energy (ATP) is available. Figure 4-4 shows the light-dependent and light-independent reactions.

Figure 4-4 The two stages of photosynthesis: the light-dependent and light-independent reactions.

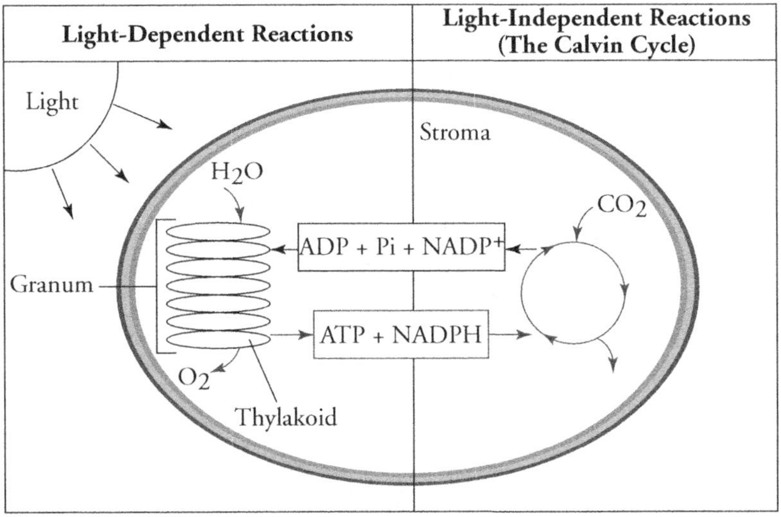

To summarize photosynthesis, energy from the sun, along with water and carbon dioxide, are used to create high-energy sugars and oxygen. The light-dependent and light-independent reactions are the processes that together allow food to be made for the plant, which in turn becomes food for us, and oxygen is released into the environment. A pretty important process for all living things!

Cellular respiration

Aerobic vs. anaerobic

All organisms must have energy to carry out life processes. During photosynthesis, plants (autotrophs) use energy from the sun to make their own food. Animals (heterotrophs) cannot sit outside and absorb the sunlight and be fed. Instead, they have to take the food that was made by the plants and break it down to produce ATP. The process of **cellular respiration** is how animals release the chemical energy from sugars to make the energy they need to grow, reproduce, and survive. Cellular respiration must have oxygen; the formula is found in Figure 4-5.

Figure 4-5 The formula for cellular respiration.

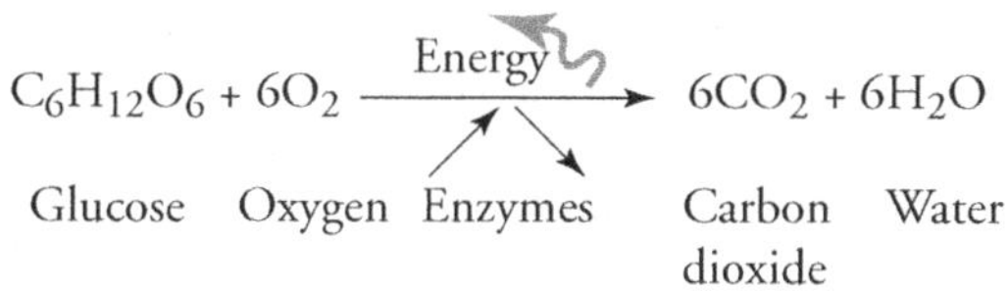

Glucose (or sugar) and oxygen are used to produce carbon dioxide and water. The organism can use the resulting energy to grow, reproduce, and survive.

If you recall, photosynthesis had two main parts: the light-dependent and light-independent reactions. Cellular respiration has three stages: glycolysis, the Krebs cycle, and the electron-transport chain. We will break down each of these three into the steps they contain.

Glycolysis is the process that gradually breaks down one molecule of glucose into two molecules of pyruvic acid. The energy released during this process is used to also create two molecules of ATP and two molecules of nicotinamide adenine dinucleotide (NADH). Glycolysis means "splitting sugars"; a six-carbon sugar, glucose, is split into two molecules of a three-carbon sugar. Glycolysis takes place in the cytoplasm of the cell and can occur **anaerobically** (without oxygen) or **aerobically** (in the presence of oxygen). As the bonds of glucose are broken down, energy is released. Although it takes two ATP molecules to get the process started, in the end glycolysis produces four ATP molecules from each molecule of glucose. Therefore, in glycolysis there is a net gain of two ATP molecules.

Fermentation is what happens when glycolysis occurs anaerobically. Have you ever been exercising and your muscles started to burn? When you exercise strenuously, your muscle cells use up all their oxygen supply and begin to convert glucose to pyruvic acid. Then an enzyme in the muscle cells converts the pyruvic acid to lactic acid. The lactic acid produces the burn you feel.

The **Krebs (citric acid) cycle** is the next step after glycolysis has split the six-carbon sugar into two three-carbon molecules called pyruvic acid. During the Krebs cycle, the pyruvic acid that was produced in glycolysis is further broken down into carbon dioxide and energy. The Krebs cycle occurs in the mitochondrial matrix and yields two more molecules of ATP.

The final step, the **electron-transport chain,** is where the majority of energy (ATP) is produced; 34 ATP molecules are made from each molecule of glucose. The details of this process are beyond the scope of this text. The electron-transport chain occurs only in the presence of oxygen; as ATP is made, water is produced.

Figure 4-6 summarizes how many molecules of ATP are made during each process of cellular respiration.

Figure 4-6 During each of the three steps in cellular respiration, ATP (energy) is produced. You can see in this chart that 2 ATP molecules are formed during glycolysis and the Krebs cycle, and 34 molecules of ATP are made during the electron-transport chain.

Energy	
Glycolysis	2 ATP
Krebs cycle	2 ATP
Electron-transport chain	34 ATP
Total	38 (36)* ATP

*36 because 2 ATP are used to transport products of glycolysis into the mitochondria.

Animal Systems

Body systems

The human body has 11 different systems that interact together to keep us alive (see Figure 4-7). In this section, we will review each system and what it does for us. Keep in mind that the important part is how these systems interact with one another. In other words, no system works alone in your body; you need all 11 systems functioning properly or you can get sick.

Figure 4-7 The 11 systems of the human body.

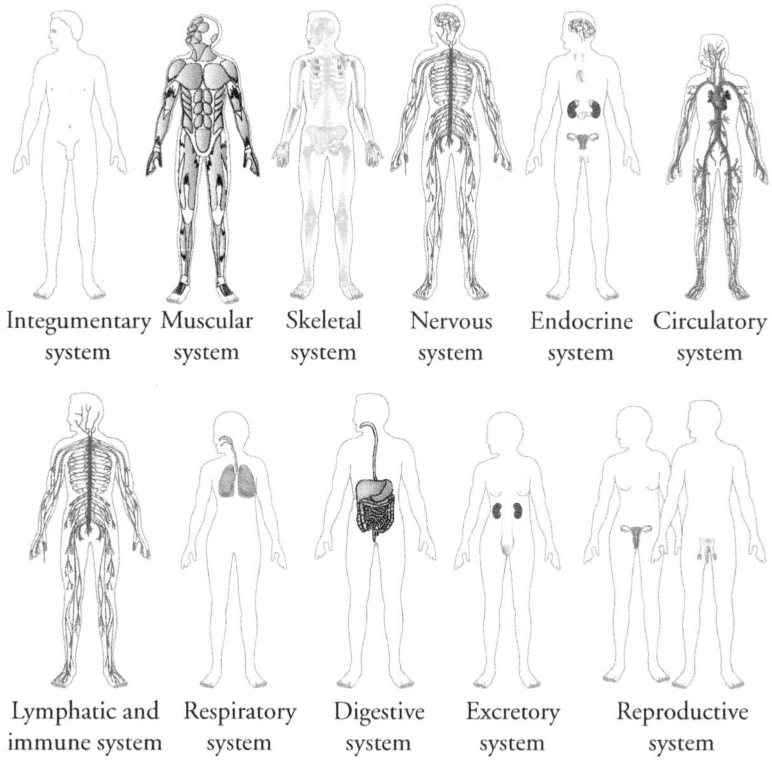

The **integumentary system** includes the largest organ in the body, the skin. The integumentary system also is made up of hair, nails, and sweat and oil glands. The main function of the integumentary system is protection from the outside and prevention of water loss.

The **muscular system** includes the three types of muscles in the body: smooth, skeletal, and cardiac. Smooth muscle makes up the organs, skeletal muscle makes up the muscles used to move the body, and cardiac muscle makes up the heart. Some muscles are voluntary, meaning we can think about them and make them move, like the muscles in the arms and legs; other muscles are involuntary, meaning they move without our thinking about it, like the muscles of the heart.

The **skeletal system** includes the bones, joints, and ligaments that keep the body moving and support and maintain the body's shape. The skeletal system is also responsible for making the red blood cells and for protecting parts of the body such as the skull, which protects the brain. The bones store calcium and phosphorous and are a place where the muscles attach to allow us to move.

The **nervous system** is extremely complex and includes the brain, spinal cord, and nerves. The nervous system is what controls all the coordination of your body's activities, and it interacts with all other systems to keep you alive. Sight, hearing, balance, touch, smell, and taste are all functions of your nervous system. Your nervous system also tells your muscles in your heart to beat, tells you to breathe without thinking about it, and allows your digestive system to break down your food. Your nervous system is what regulates your body's response to both internal and external stimuli.

The **endocrine system** includes hormones and glands and releases the hormones into the blood to be used by the body. The endocrine system helps to control the body's growth and development, helps us to respond to outside stimuli from the environment (for example, sweating when we get hot), and controls things in the body like blood sugar levels.

The **cardiovascular (circulatory) system** includes the heart, blood vessels (arteries, capillaries, veins), and blood. The function of the circulatory system is to transport substances around the body. This includes carrying oxygen from the lungs to the body, bringing nutrients to the cells, transporting hormones, and getting rid of metabolic waste products such as carbon dioxide.

The **immune system** (which includes the lymphatic system) is the way the body protects us from infections and disease. The immune system includes cells known as B cells and T cells to destroy pathogens and uses the body's circulatory and lymphatic systems to move about.

The **respiratory system** is the body's way of taking oxygen in the atmosphere and carbon dioxide in the blood and exchanging them from the body to the environment (and vice versa). The respiratory system includes the nose, sinuses, pharynx, larynx, trachea, and lungs.

The **digestive system** takes the food we eat and breaks it down into smaller pieces that can be used by the body. It is also responsible for helping to get rid of undigested food. The parts of the digestive system include the mouth, esophagus, stomach, and small and large intestines. The digestive system uses enzymes and mechanical means to break down food.

The **excretory system** (which includes the urinary system) removes metabolic wastes from the body and includes the lungs, skin, kidneys, ureters, urinary bladder, and urethra.

The **reproductive system** consists of the body's sex organs and is responsible for creating offspring for a species. Some of the many organs in the reproductive system include the penis, testes, scrotum, vas deferens, ovaries, vagina, and uterus. The reproductive systems of males and females work together for the purpose of sexual reproduction.

Enzymes

The next biological process we will discuss is the role of enzymes in living things. Many chemical reactions that are important to life would happen very slowly if it were not for enzymes. **Enzymes** speed up chemical reactions in living things. A **catalyst** is something that speeds up the rate of a chemical reaction, and an enzyme is a type of biological catalyst made of protein. All cells contain enzymes.

Enzymes are essential to life. After catalyzing their reactions, enzymes are not used up and therefore can be used over and over again. You have enzymes that break down food, enzymes that remove carbon dioxide from your blood, and enzymes that are important in your immune system.

Enzymes are also very specific. Oftentimes the function of enzymes is described as the "lock and key" model. Each enzyme is specific to the substrate it binds with. The enzyme fits together with its substrate like a key fits a lock. The specific location on the enzyme that comes into contact with the substrate is called the enzyme's **active site.** Have you ever known someone who cannot drink milk because he gets sick if he does? The specific enzyme lactase can break down the sugar lactose found in milk and other dairy products. If a person's body does not make enough of the enzyme lactase, he can't break down lactose and gets sick (diarrhea, bloating, gas, and so on) if he ingests milk products. Look at Figure 4-8 of an enzyme and its very specific substrate.

Figure 4-8 An enzyme is very specific to the substrate it binds with. Much like a lock and key, only certain enzymes bind to certain substrates.

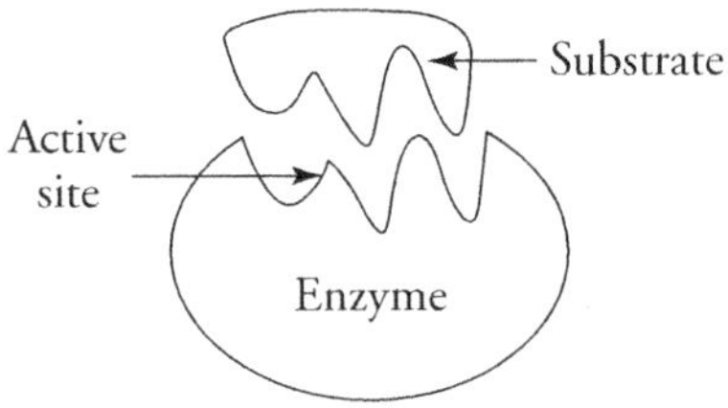

Enzymes can break down substances or combine more than one molecule together to make something new. An enzyme breaks or builds chemical bonds, and when it is finished, it goes on to do it again and again. Figure 4-9 shows a picture of an enzyme breaking down or making something new.

Figure 4-9 An enzyme breaking down a substrate or combining two things to make something new.

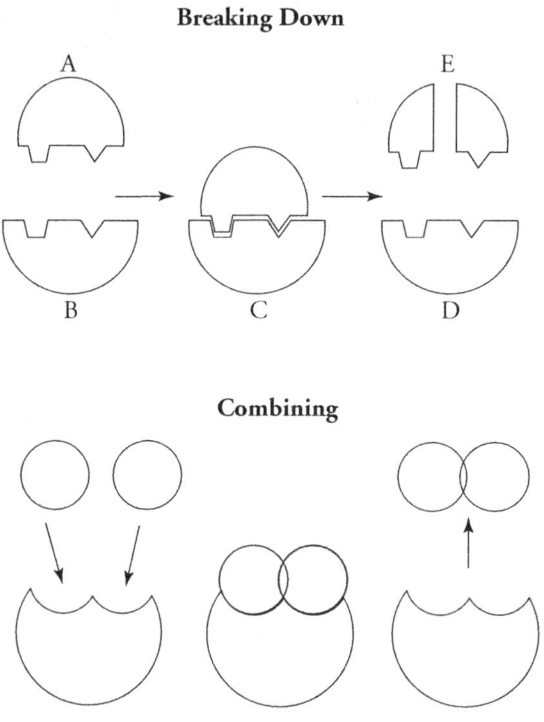

Homeostasis

Homeostasis is the body's way of regulating and maintaining the constant internal condition needed for life. Our bodies have numerous ways of maintaining homeostasis. Below is a chart of a few examples. Think of homeostasis as your body's way of returning you back to normal.

Examples of Homeostasis
Blood sugar levels are controlled by insulin.
Body temperature is regulated by sweating and shivering.
The body's water content is controlled by the lungs, skin, and urine production.
The heart senses changes in blood pressure, telling the heart to slow down or speed up.
If bacteria or a virus gets into the body, the immune and lymphatic systems help to fight it.
The thyroid helps maintain proper calcium levels in the blood.

Levels of organization in the body

There are five levels of organization in animal systems: cells, tissues, organs, organ systems, and organism. The smallest unit of life is the **cell,** and each type of cell is particular to the function it performs for the body. **Tissues** are a group of similar cells that work together to carry out a particular function. **Organs** are then groups of tissues that perform a function in the body. For example, the heart, lungs, and skin are all organs of the human body. An **organ system** is then multiple organs that work together. For example, the digestive system includes the stomach, intestines, and so on. Finally, all the organ systems make up the **organism** (for example, a human). Figure 4-10 shows the hierarchy of levels of organization in living things.

Figure 4-10 Five levels of organization in living things.

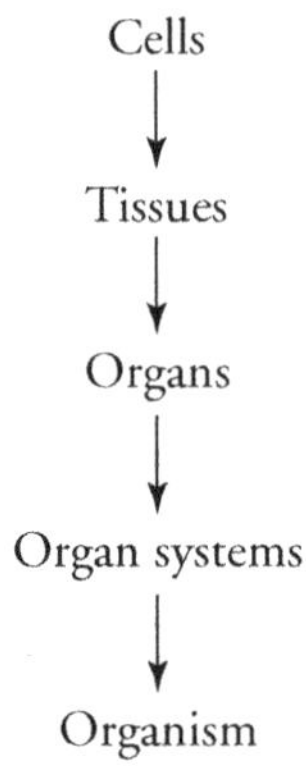

Plant Systems

Transportation

Plants have tissues that can carry water and nutrients. The roots take up water from the soil and need to bring the water to the rest of the plant. They do this through specialized vascular tissue called **xylem.** Plants also need to take the food (dissolved sugars and other organic compounds) that is made in the leaves and distribute it throughout the plant. Plants do this though specialized vascular tissue called **phloem.** Figure 4-11 shows the difference between the two types of vascular tissues. In summary, phloem moves food (glucose) and xylem moves water for the plant.

Figure 4-11 The difference between xylem and phloem vessels.

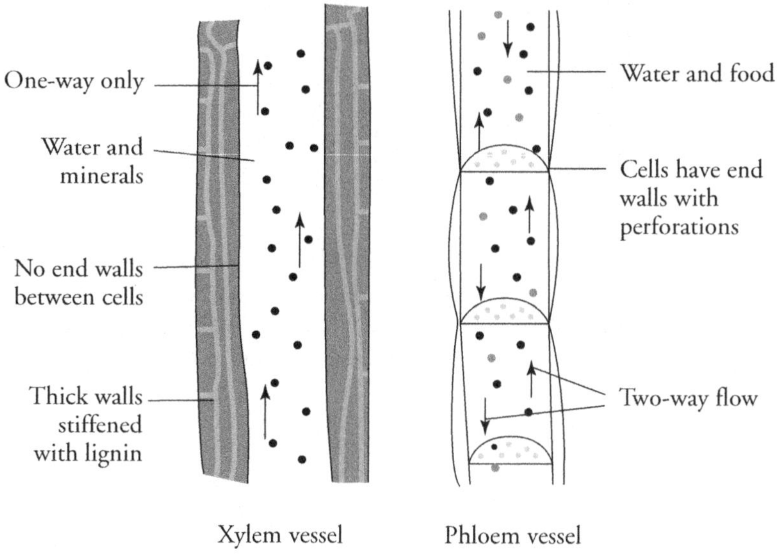

Xylem and phloem are typically found in the stem of the plant, but can be found in other places as well. Xylem and phloem in a plant are like veins and arteries in an animal. Figure 4-12 shows a cross-section of a stem with xylem and phloem labeled.

Figure 4-12 Xylem and phloem in a plant stem.

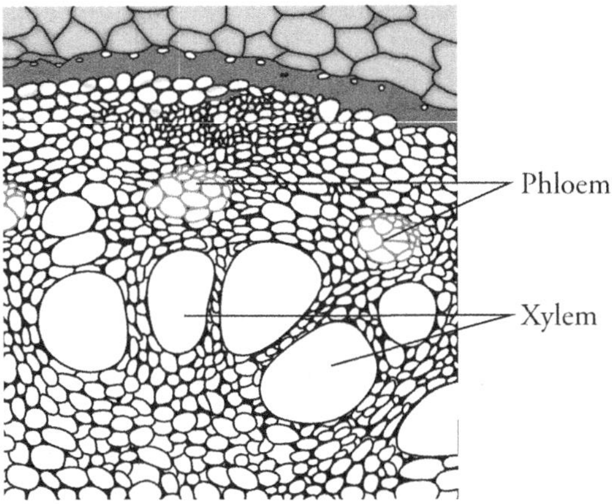

Reproduction

Plants can reproduce both sexually and asexually. An example of asexual reproduction would be when a plant puts out runners, side shoots, bulbs, or tubers (see Figure 4-13). The new plant is identical to the parent plant.

Figure 4-13 Asexual reproduction.

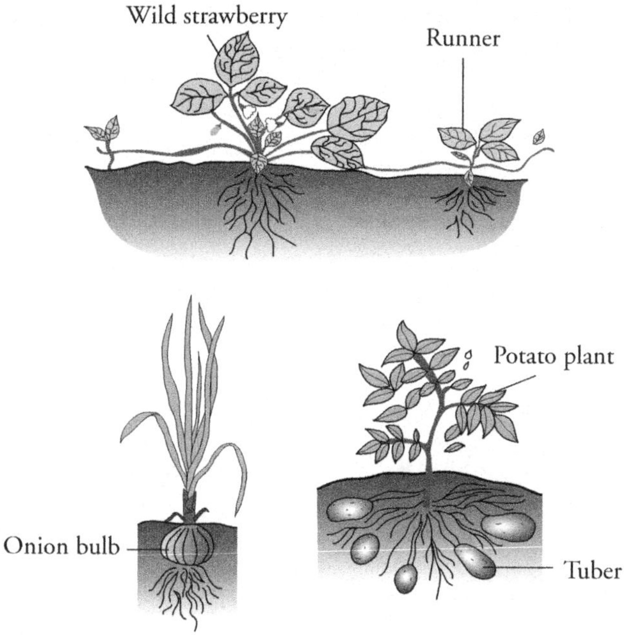

Other plants reproduce sexually with sperm and eggs. There are several ways this occurs. For example, once a moss plant is fertilized, a stalk forms with a capsule at the top. The capsule matures and releases spores that burst free from the plant and form a new moss plant.

Another way is with conifers or cone-bearing plants. If you look at a pine tree, you will generally see two types of cones. The female part is usually lower on the branches, and the male part is higher, in the top of the trees. The female part of the plant has a cone that opens up with an ovule on each scale of the pine cone. The male pine cone will open in the spring and release hundreds of pollen grains (which contain the sperm) that can land on the female pine cone, allowing the sperm to reach the egg. These fertilized eggs then develop into the seeds that are released to the environment. Figure 4-14 depicts the male and female cones of a conifer.

Figure 4-14 A conifer (cone-bearing) tree.

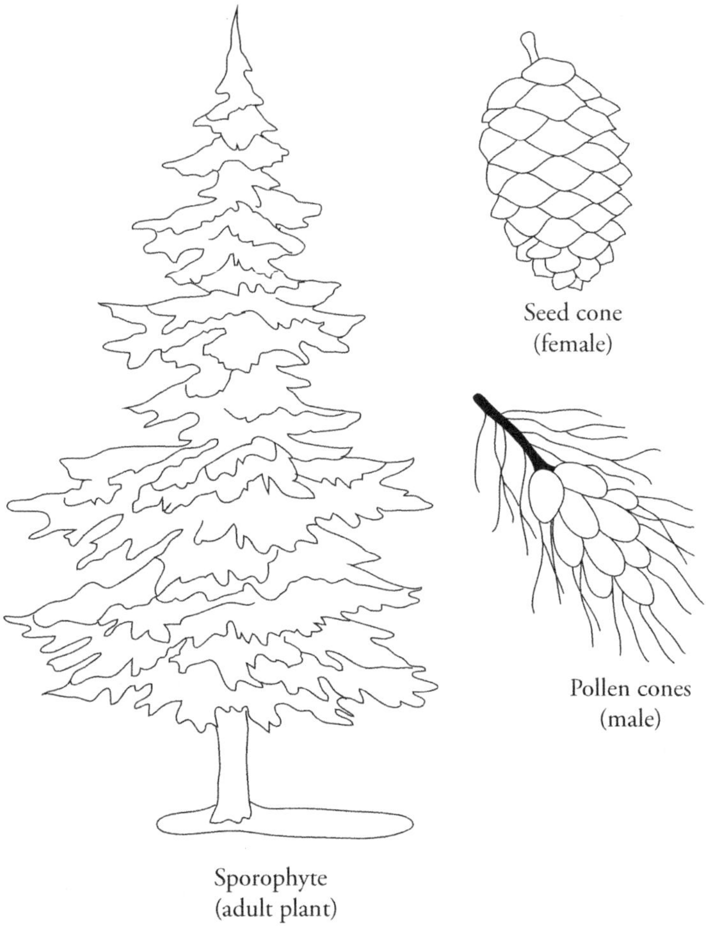

The final type of sexual reproduction we will discuss is in flowering plants. Flowers have many organs to help with reproduction. Both male and female parts can be found on a single flower, and a flower is often fertilized by birds, insects, wind, and other animals. Figure 4-15 shows the parts of a flower.

Figure 4-15 A typical flower.

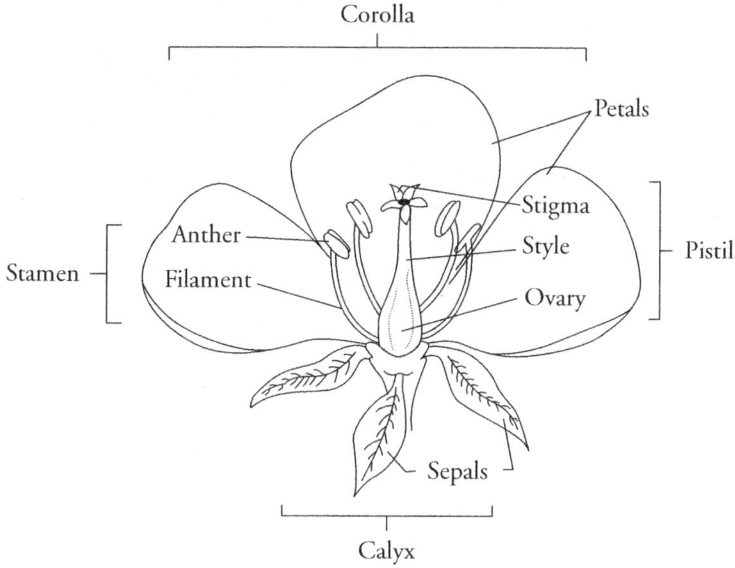

Locate the male parts and the female parts in Figure 4-15. The male parts include the stamen, anther, and filament. The female parts include the pistil, stigma, style, and ovary. Let's break down each structure into its function.

Stamen: The male structure of the flower

Anther: The structures where the pollen grains are produced

Filament: The stalk that holds the anther

Pistil: The female structure of the flower

Stigma: The sticky structure that captures the pollen

Style: The stalk that holds the stigma

Ovary: Part of the female reproductive structure of the flower; where the new plant begins to form

Response

Just like animals, plants must maintain homeostasis. One way this can occur is through response to their external environment. For example, plants have specialized structures in their leaves called **stomata** (singular, **stoma**). The carbon dioxide that is needed for photosynthesis enters through the stomata. These structures can open and close depending on weather conditions. If it is hot and dry outside, the stomata will close to prevent **transpiration,** the process of water evaporating from the plant structures like the leaves. See Figure 4-16 for a picture of a leaf and the stomata.

Figure 4-16 Stomata.

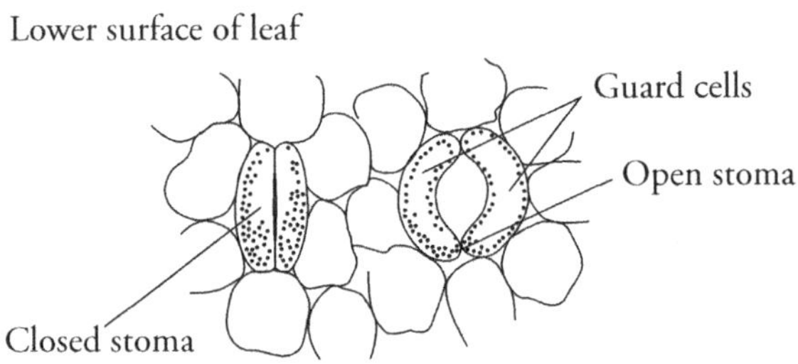

Another way plants respond to their environment is by moving toward the light. This process, known as **phototropism,** allows the plant to bend toward the light source so that it can capture the most light for photosynthesis. See Figure 4-17 for an example.

Figure 4-17 Phototropism.

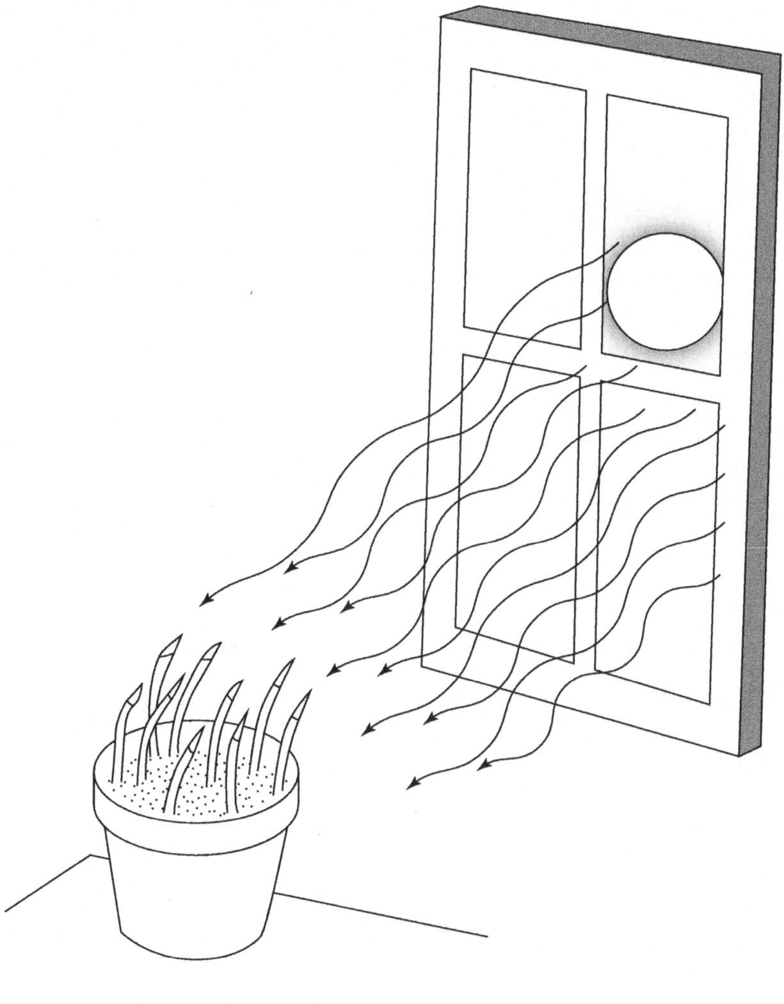

Other responses in plants are **thigmotropism,** which is how vines are able to find an object like a fence or tree and wrap around it, and **gravitropism,** which is a plant's response to gravity. The roots will always grow downward toward gravity (positive gravitropism) and the stem grows upward against gravity (negative gravitropism).

Chapter Check-Out

1. When you are exposed to a virus or bacteria that could potentially make you ill, your body systems work together to protect you. Which two of your body systems would do this?
   a. circulatory and immune
   b. digestive and endocrine
   c. circulatory and reproductive
   d. integumentary and muscular

2. What is the difference between cellular respiration and photosynthesis?
   a. Cellular respiration requires sunlight.
   b. In photosynthesis, carbon dioxide is released and oxygen is absorbed.
   c. Cellular respiration releases energy in a step-wise manner as ATP molecules.
   d. In photosynthesis, glucose is broken down into water and carbon dioxide.

3. A student performs an experiment where she puts a plant by a window and observes the plant each day for a week. Below is a picture of the plant on day 1 and day 5.

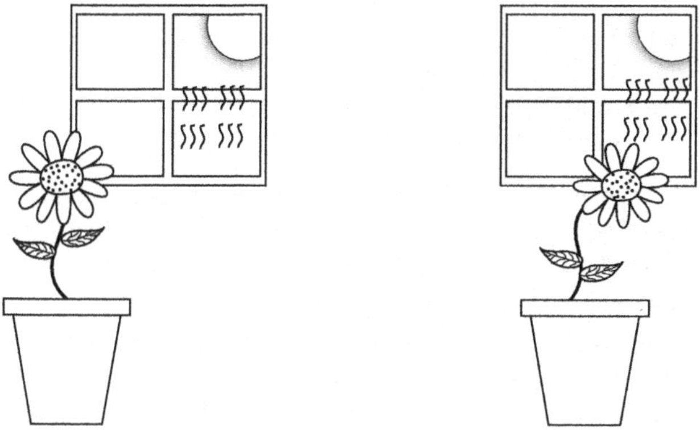

Day 1 Day 5

At the end of the week, the student makes observations and concludes that the reason the plant has moved in the direction of the window is because of

a. thigmotropism.
b. gravitropism.
c. phototropism.
d. transpiration.

4. Some people do not have the ability to digest milk. If they consume milk products, they will become sick and can have an upset stomach, diarrhea, or other digestive problems. They cannot digest milk because

a. enzymes are not specific, so their bodies are using the enzyme to break down a different substance they consumed.
b. they do not have enough of the enzyme lactase to break down the lactose in the milk.
c. enzymes can be used up, and their bodies are slow to generate more of the enzyme, so they become ill.
d. enzymes are carbohydrates, and they are not consuming enough carbohydrates in their diets.

5. In your body, your integumentary system provides protection and prevents water loss, while your respiratory system exchanges gases (oxygen and carbon dioxide) with the environment. This would be similar in a plant to what two parts?
   a. xylem and phloem
   b. guard cells and stomata
   c. stamen and pistil
   d. anther and filament

6. A runner is preparing for a marathon race. During the marathon, the runner's body systems work together to keep his body in homeostasis. Which of the following correctly describes an interaction that occurs between two body systems of the runner?
   a. The endocrine system secretes hormones that tell the runner's muscular system how to react.
   b. The integumentary system works with the digestive system to increase the rate of digestion so that the runner can make the nutrients he consumed last longer.
   c. The excretory system tells the skeletal system to absorb calcium for increased skeletal strength.
   d. The circulatory system increases blood pressure so the tissues get more oxygen and the respiratory system increases the runner's breathing rate.

7. The concentration of glucose in the blood is regulated by hormones. Even after a 24-hour fast, glucose levels are reduced only very slightly. The hormone glucagon works by increasing blood sugar concentrations by freeing up glucose stored as glycogen in the liver and muscles. This ensures stable blood glucose levels, even when a person skips a meal. This process is an example of which of the following?
   a. the role of the endocrine and reproductive systems working together
   b. a feedback mechanism to maintain homeostasis
   c. the endocrine system releasing hormones so the immune system can keep the body healthy
   d. the nervous system telling the immune system how to maintain homeostasis

8. Put the following in order from largest to smallest.
   a. cell, tissue, organ, organ system, organism
   b. cell, tissue, organ system, organ, organism
   c. organism, organ system, organ, tissue, cell
   d. organ system, organism, organ, tissue, cell

9. Phosphorus is a plant nutrient that is found in the soil and is important in nucleic acids, phospholipids, coenzymes, DNA, NADP, and ATP in a plant. Phosphorus helps to decompose carbohydrates that a plant produces in photosynthesis, is used in protein synthesis, and does many other processes that a plant needs to grow and reproduce. Which statement correctly shows how the roots and shoot systems of a plant work together to allow phosphorus to get from the soil to the plant?
   a. The plant absorbs phosphorus in the roots and transports it to the plant's tissues by the shoot system.
   b. Phosphorus is absorbed in the leaves and is transported to the roots via the shoot system.
   c. The shoot system transports phosphorus to the roots in a process called transpiration.
   d. The roots absorb phosphorus and make sugars that are taken to the leaves, where the stomata release gases to the environment.

10. What reaction is occurring when solar power is used in the synthesis of ATP and NADPH in the thylakoids of chloroplasts?
    a. carbon-fixation reactions
    b. light-independent reactions
    c. light-dependent reactions
    d. fermentation reactions

Answers: 1. a **2.** c **3.** c **4.** b **5.** b **6.** d **7.** b **8.** c **9.** a **10.** c

Chapter 5

INTERDEPENDENCE WITHIN ENVIRONMENTAL SYSTEMS

Chapter Check-In

- ❑ Ecological succession on land and in water
- ❑ Biological interactions
- ❑ Energy flow in ecosystems
- ❑ Nutrient and water cycles

Ecology, often called environmental science, is the study of how organisms interact with other organisms or with their environment. This environment, or **biosphere,** is the part of the earth where life exists; it includes the land, water, air, and atmosphere.

Ecological Succession on the Land

Change of an ecosystem over time

Ecosystems are always changing over time. An **ecosystem** is a biological community, including all the non-living factors that affect it. Some changes occur because of natural processes, and others occur because of human interference. These series of changes are called **ecological succession.**

Primary succession

Primary succession (Figure 5-1) occurs where an area has no soil, such as after a volcanic eruption, or on bare rock if a glacier melts, leaving the rock exposed. The first stage of primary succession is when lichen or moss begins to break down the rock, forming a small amount of soil. Once this small amount of soil has formed, small plants such as grasses can take root, further breaking down the rock and forming more soil. Now, larger plants can begin to grow, and finally shrubs and trees will occupy the area. Primary succession is very slow and can take hundreds to thousands of years.

Figure 5-1 Primary succession.

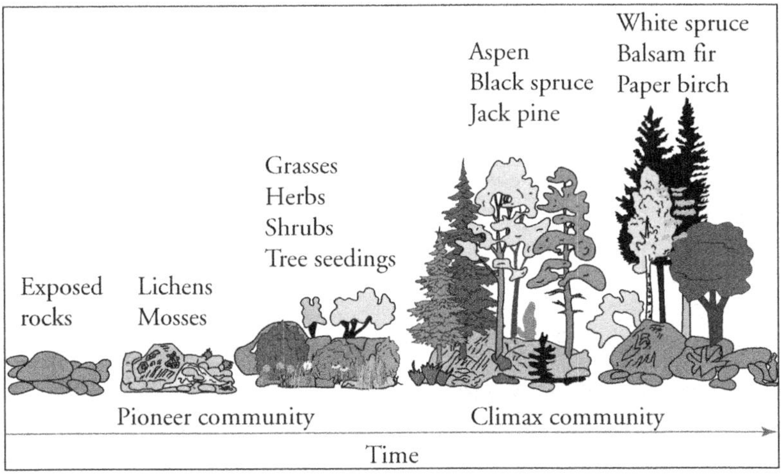

Secondary succession

Secondary succession (Figure 5-2) happens after an area that has soil has been destroyed (for example, by a forest fire). After the fire, the soil is still present, so seeds from area plants can come in and grow. The stages of secondary succession are small plants and weeds, followed by larger plants and shrubs, giving rise to small trees, which are eventually replaced by larger trees. Secondary succession generally does not take as long as primary succession because the soil is already present.

Figure 5-2 Secondary succession.

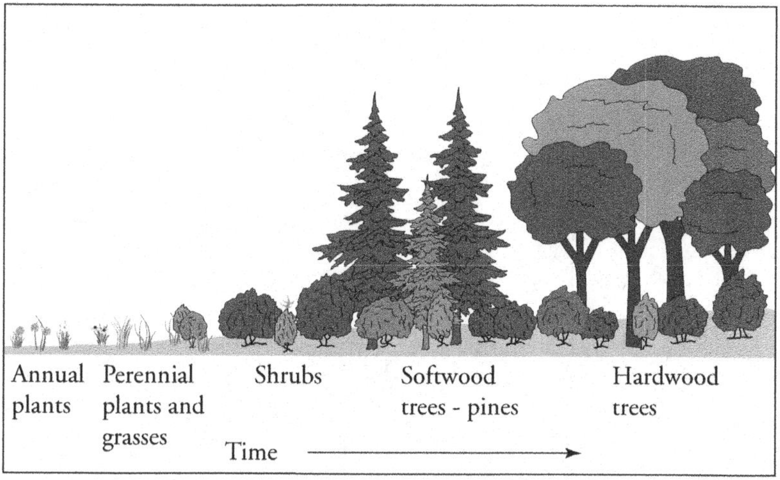

Climax community

The **climax community** exists when the area is returned to the original, mature ecosystem. In the example of the forest fire, the climax community is when the area is back to a forest.

Ecological Succession in the Water

Change of an ecosystem over time

Similar to succession on the land, succession can also occur in aquatic environments such as a pond (see Figure 5-3). Think of a lake and all the mud, leaves, and so on that slowly begin to fill in the lake. After hundreds or thousands of years, the lake will begin to get shallower. Eventually, if nothing stops this, the lake can end up completely filling in and become dry land.

Figure 5-3 Succession in a pond.

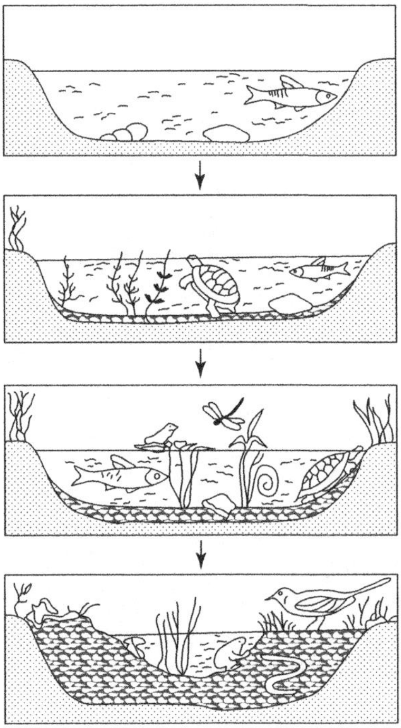

Biological Interactions

Populations

Populations are groups of individuals of the same species living in the same area. For example, a herd of caribou living in Alaska would be a population. Populations can breed with one another to produce viable offspring and interact with other members of the same population, as well as with members of other populations and the non-living world.

Communities

Communities contain many different populations of both plants and animals that interact together. A forest would be an example of a place containing many different communities. A community might be the trees, grasses, deer, squirrels, and so on living together in a forest ecosystem.

Ecosystems

The groups of living and non-living things interacting together make up an **ecosystem** (see Figure 5-4). From the example above, this would be the trees, grasses, deer, and squirrels, along with the rocks, water, and air in the area. Living things are known as **biotic,** and non-living things are called **abiotic.**

Figure 5-4 Components of an ecosystem.

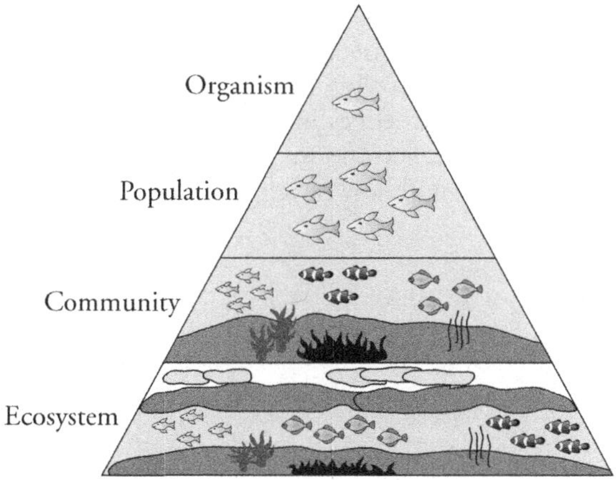

How environmental change affects ecosystems (HIPPCO)

A change to an ecosystem that has been caused by humans is known as an **anthropogenic** change. Humans can affect ecosystems both positively and negatively. You may know or have been a part of a positive change to the environment. Maybe you helped to clean up litter along the edge of a river or you planted plants after a hurricane; these would be positive changes. However, many anthropogenic changes are negative. One way to remember the major negative changes that humans can do is to remember the mnemonic **HIPPCO.**

Habitat loss
Invasive (introduced) species
Pollution
Population overcrowding
Climate change
Overharvesting

Habitat loss is the first negative consequence that humans are doing to the environment and is considered the most damaging of all. As humans do things such as build larger cities, mine for minerals, cut down forests, and build dams on major rivers, they are damaging the natural environment where many communities live. This destroys the habitat of the species that live there. Humans are also taking species' habitats and causing habitat fragmentation. **Habitat fragmentation** (Figure 5-5) is when human development splits ecosystems into many different pieces. One example could be a road that cuts through an existing environment. Now animals can have a problem trying to migrate or mate. In addition, now you increase the number of edges in the habitat, which causes an **edge effect.** Increased edges are harmful to some species that need to be in the middle of a dense forest and now have more edge habitat to deal with.

Figure 5-5 Habitat fragmentation.

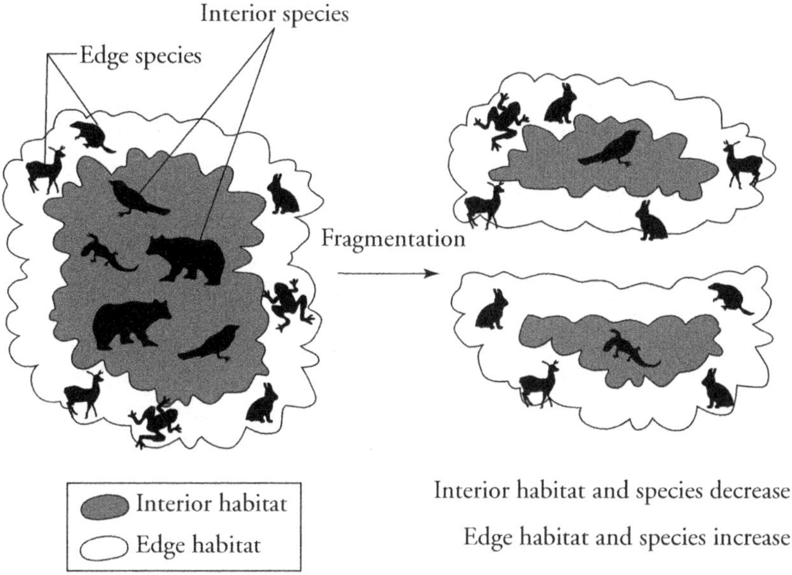

Invasive (introduced) species are the next negative impact that humans have on the environment. An invasive species is a species that is brought into an environment by humans where it does not naturally live. Humans have introduced invasive species accidently and on purpose. An invasive species could be a plant that someone saw in another country and brought into a new environment; in the case of the fire ants of North America, the ants hitchhiked on a boat where they had made a nest in a wooden palate. Invasive species may not harm an ecosystem at all, but many times the invasive species are able to outcompete the native species, causing an impact on the biodiversity of the area. Invasive species can cause millions of dollars in economic damage as well.

Pollution is another negative impact humans have on the environment. Pollution can change the air, water, and soil, which can harm the biodiversity in the area. One of the biggest problems for biodiversity is when a pollutant biologically magnifies in an environment. **Biological magnification** is when concentrations of a harmful substance increase in organisms as you move up the food chain or food web. Dichloro-diphenyl-trichloroethane (DDT) is an example of a chemical that can biomagnify. DDT is a pesticide that was used from the 1940s to the 1970s to kill disease-carrying insects. At the time, it seemed like the perfect pesticide. It was relatively inexpensive and worked very effectively. What was unknown at the time was how DDT would work its way up the food chain to toxic levels (see Figure 5-6).

Here is how it would work. The DDT made its way into the water, where it was taken in by the phytoplankton. These phytoplankton that each had a little DDT were eaten by the zooplankton. The zooplankton ate enough phytoplankton to get more DDT in their bodies. The zooplankton were then eaten by small fish, which eat a lot of zooplankton. Now the concentration in the small fish is even greater. Finally, the fish-eating birds like ospreys or bald eagles, which again eat many small fish, would accumulate an even higher amount of DDT in their bodies. Interestingly enough, the fish-eating birds did not die, but when they went to lay eggs, the eggs did not have viable shells. This caused the shells to be fragile, which led to the death of the baby birds. It doesn't take long for a species without viable offspring to decrease in numbers. In some cases, the birds ended up on the endangered species list. DDT was banned in many parts of the world after the environmental impacts of the chemical were discovered.

Figure 5-6 DDT biological magnification.

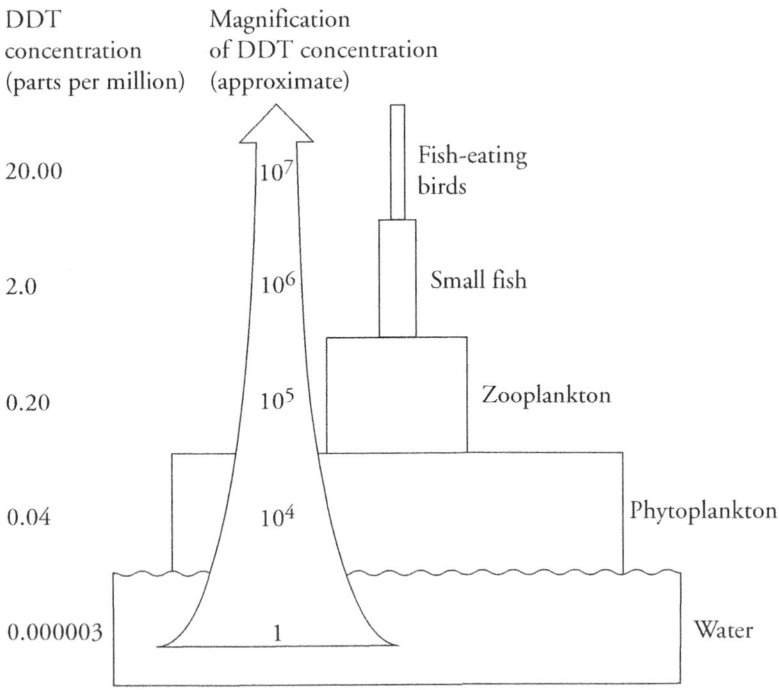

Population overcrowding is another human-caused negative impact on the environment. There are more than 7 billion people on Earth, each having an environmental impact. The human population is increasing exponentially, and at the current rate it will likely double in less than 60 years. Humans and wildlife do not often mix, and wildlife most often loses when they do. As the human population continues to grow, we will demand more space for development, will demand wildlife products, will pollute, and will continue to harm the animals and plants we come into contact with.

Climate change is next. Earth is becoming hotter. Over the past 100 years, Earth's temperature has increased by 1° Fahrenheit. That may not seem like much to you, but you must remember that it is not 1° everywhere. It is an average of 1°. Many places are getting much hotter, including the poles, where the majority of the ice on our planet is found. This increase in temperature is melting the ice caps and glaciers, causing animals like the polar bear to lose precious habitat; it is also changing weather patterns across the globe. Some impacts of climate change include stronger storms like hurricanes, droughts and floods, heatstroke in humans, rising sea levels due to the melting of the ice and thermoexpansion of the oceans, and flooding of coastal ecosystems. Climate change is thought to be occurring because of humans burning fossil fuels (coal, oil, natural gas), adding carbon dioxide to our atmosphere. Carbon dioxide is a greenhouse gas that can trap heat. We are also raising cattle that emit methane, another greenhouse gas.

Overharvesting is the last in our mnemonic list of ways humans are harming the environment. Overharvesting can be overfishing, overhunting, and so on. Any time we take too many of an organism for our benefit, we are overharvesting. Legislation has helped this problem; laws have been enacted to limit the number or amount of animals/plants that can be harvested.

Symbiotic relationships

Many species have adapted to have a relationship with other species. **Symbiosis** is the interactions between two living things that live together. There are three different kinds of symbiosis: mutualism, commensalism, and parasitism.

Mutualism is when two species live closely together and both benefit from the relationship. An example of mutualism would be a clown fish and a sea anemone. The clown fish uses the anemone for shelter, and the anemone gains waste from the clown fish for nutrients.

Commensalism is when one species benefits and the other is neither helped nor harmed. An example of this would be lichens that grow on trees. The lichens benefit by getting up higher in the tree so they can get light. The tree, on the other hand, is not really impacted by the lichen.

Parasitism is when one species benefits and the other is harmed. An example of this would be a tapeworm in a dog. The tapeworm gets nutrients from the dog, but the dog is harmed because the tapeworm is taking his nutrients. Another example would be a flea on a dog.

A final type of relationship that exists are feeding relationships. One example is a **predator-prey** relationship. The predator is the animal that consumes the prey. An example would be a bobcat and a deer. The bobcat is the predator, and the deer is the prey.

Adaptations of organisms in different ecosystems

An **adaptation** is a characteristic that an organism has inherited that enables it to survive in its environment; the organism then passes these genes to its offspring. For example, a plant that lives on the floor of the rain forest may have developed large leaves over time. These large leaves allow the plant to get light in a place that doesn't have much light. The floor of the rain forest is dimly lit due to the large trees that occupy the rain forest canopy, blocking most of the light. On the other hand, plants in the desert have adapted to be able to store water in modified structures for long periods of time. It may be months between rainfalls in the desert, and the plants must be able to store water that they can use slowly when rainfall is not around.

Homeostasis is the ability of organisms to prevent fluctuations in their internal environment so that they can live. When you go on a run, your body temperature increases. This could be dangerous to you if it were left unchecked. Your body maintains homeostasis by sweating. Sweat cools you off and returns your body to a temperature that is safe. Another example is the fever you may suffer when you get sick. Bacteria in your body can survive in certain temperatures. When your body gets sick, you begin to run a fever. Your body is trying to increase the temperature to kill the bacteria that are making you sick. This is your body's way of maintaining homeostasis.

Energy Flow in Ecosystems

The flow of energy in an ecosystem is one of the most important factors of the ecosystem. Energy is needed by all organisms in order to survive. The sun is the main energy source for life on Earth. Some organisms can make their own food, while others must eat other organisms. Scientists use simplified representations to demonstrate the flow of energy in ecosystems.

Producers and consumers

A **producer** is an organism that can make its own food. Plants, some algae, and certain bacteria are able to use light from the sun to make their own food or capture energy out of chemicals. Another name for a producer is an **autotroph.** *Auto* comes from the Greek word meaning "self," and *troph* is Greek for "nourishment."

A **consumer** is an organism that must get its energy from eating other living things. Animals, fungi, and some bacteria are consumers. Another name for a consumer is a **heterotroph.** *Hetero* comes from the Greek word meaning "different," and as we mentioned earlier, *troph* is Greek for "nourishment."

A **primary consumer** is an organism that eats a plant, or producer; a **secondary consumer** is an animal that eats a primary consumer; and a **tertiary consumer** is an animal that eats a secondary consumer.

A heterotroph that eats only plants is called a **herbivore,** while one that eats other animals is a **carnivore. Omnivores** are animals that eat both plants and other animals. Finally, an animal that eats dead material in an ecosystem is a **detritivore.**

Food chains and food webs

A simple way to look at the flow of energy in an ecosystem is through a **food chain.** A food chain shows the relationship between species and how the energy flows in the system. Figure 5-7 shows the food chain of an owl. Notice the direction of the arrows. The arrows always point from the food item to the organism that eats it. The plant is eaten by the insect, so the arrow points from the plant to the insect. The plant is going into the body of the insect. The insect is going into the body of the mouse so the arrow points from the insect to the mouse, and so on. This simple diagram shows how energy flows in an environment.

Figure 5-7 The food chain of an owl. A food chain shows the path of energy from one living thing to another.

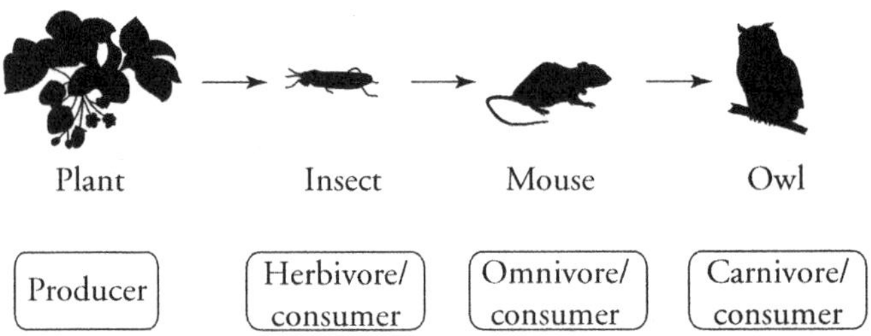

A more realistic version of how energy flows is a **food web.** A food web shows a number of different food sources for an animal. In Figure 5-7, insects are not the only thing that mice eat. Mice eat other organisms like seeds from plants. We need a way to show the multiple ways an organism can transfer energy.

In Figure 5-8, you can see that the mouse might get eaten by the wild cat, the owl, or the snake. You can also see that the lion might eat the goat, the jackal, or the wild cat. The arrows still point from the food to the animal that eats it, but it is a more complex way to show energy flow.

Figure 5-8 Forest food web.

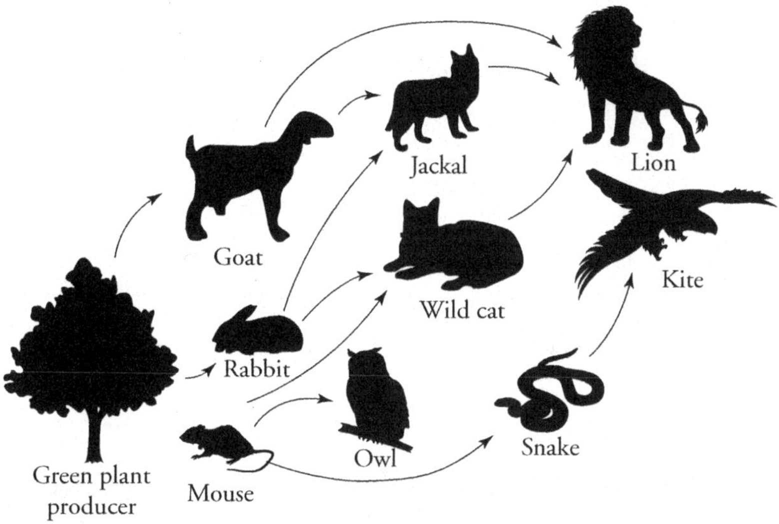

In both a food chain and a food web, we are missing a very important part of the ecosystem: the detritivores/decomposers. These organisms break down dead or dying materials and keep the entire food chain/web going by recycling the nutrients back into the environment.

Another way to show the transfer of energy in an ecosystem is with an **energy pyramid.** An energy pyramid shows the amount of energy that is available at each level of the ecosystem. Let's look back at our food chain (Figure 5-7). We had an owl that ate a mouse that ate an insect that ate a plant. Let's assume that the plant had 10,000 kilocalories of energy available and when the insect eats the plant it gets all 10,000 kilocalories. But the insect then uses that energy to move, jump, reproduce, and live. We assume that the insect uses 90 percent of the energy that it obtained from the plant and stores 10 percent; this 10 percent (1,000 kilocalories) is what's available for the next level of consumer. When the mouse eats the insect, it gets 1,000 kilocalories. The mouse uses 90 percent of this energy to move, jump, reproduce, and live, leaving only 10 percent, or 100 kilocalories. Now the owl comes around and eats the mouse. So, the owl gets 100 of the original 10,000 kilocalories from the plant. After the owl uses 90 percent of the energy, 10 kilocalories are still available. Below is a diagram of the **10 percent rule** (see Figure 5-9).

Figure 5-9 The 10 percent rule.

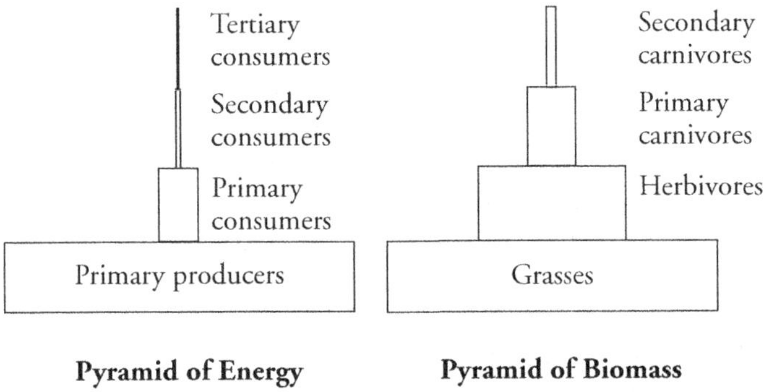

This 10 percent rule also helps to explain why there are fewer top predators (tertiary consumers) in an ecosystem than primary or secondary consumers. There is just not enough energy around to support them.

So, where does the energy go? It is used up by the animal mainly in the form of heat, where it returns to the ecosystem.

Nutrient and Water Cycles

The carbon cycle

Carbon is found in the air, in the water, and on land. All living organisms are made up of carbon, and carbon cycles around the planet in many ways. For example, when an organism dies and decomposes, carbon is returned back to the land. When an animal, like a human or deer, exhales, carbon is released into the atmosphere in the form of carbon dioxide. When we burn fossil fuels such as coal, oil, or natural gas, we add carbon to the atmosphere. This carbon formed these fossil fuels millions of years ago as organisms died and decomposed and were put under pressure. Rocks such as limestone and dead organic matter found in the soil are also sources of carbon. Figure 5-10 shows a simplistic view of the **carbon cycle.**

Figure 5-10 Carbon cycle.

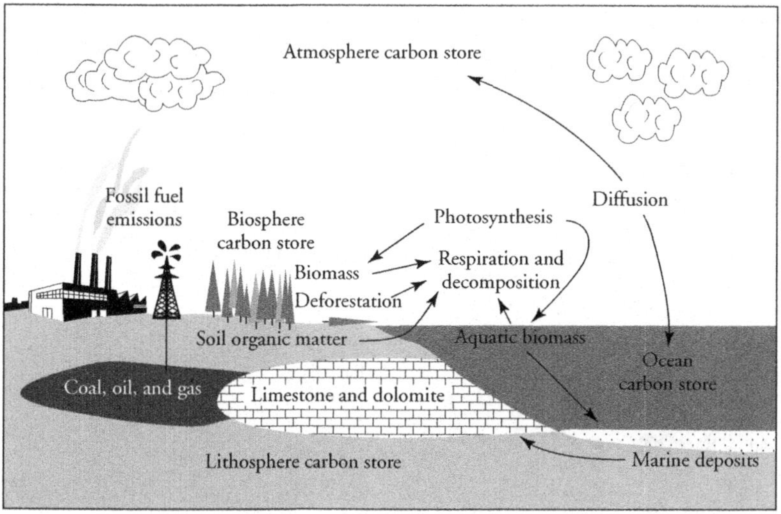

Anywhere that carbon is stored is known as a **carbon sink.** A tree that takes the carbon dioxide from the atmosphere and converts it into glucose, stores the excess carbon and is therefore a carbon sink. The ocean is another huge carbon sink. Phytoplankton, algae, and aquatic plants are able to take the carbon that diffuses into the ocean, convert it into glucose, and store the carbon in their tissues. Photosynthesis is the main process that reduces the amount of atmospheric carbon dioxide.

One concern with burning fossil fuels is that we are taking carbon that has been buried under the ocean or land for millions of years (old carbon) and returning it to the atmosphere to cycle today (new carbon). This additional atmospheric carbon is then a major contributor to the climate change we discussed earlier in this chapter.

The nitrogen cycle

Nitrogen makes up about 78 percent of our atmosphere. However, we cannot breathe in nitrogen and use it as we can with oxygen. Instead, we must obtain nitrogen in the food we eat. The nitrogen cycle is shown in Figure 5-11.

Figure 5-11 Nitrogen cycle.

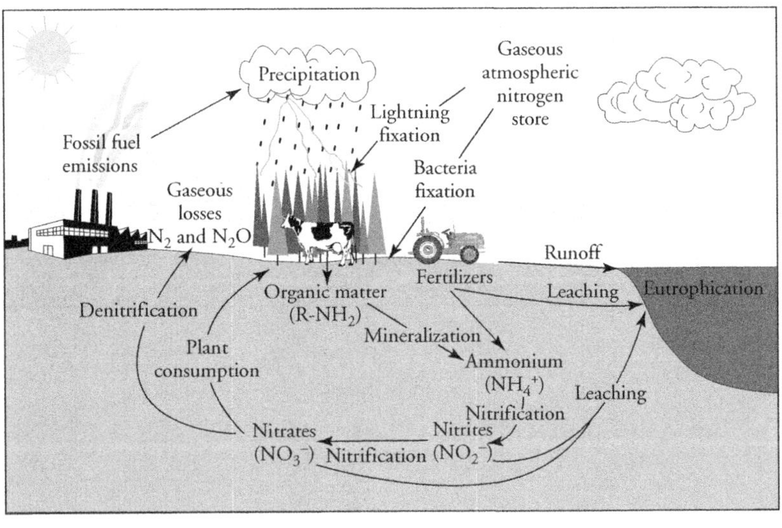

Only certain bacteria that live in the soil and in the roots of some plants are able to take atmospheric nitrogen and convert it into a usable form. This process is known as **nitrogen fixation.** In nitrogen fixation, the bacteria take atmospheric nitrogen and convert it to ammonia in a process known as **ammonification.** Then there are other bacteria in the soil that are able to convert the ammonia to nitrites and nitrates. This process is known as **nitrification.** Plants can now take the nitrites and nitrates and use them to build amino acids and proteins in a process known as **assimilation.** Animals can then get the nitrogen by eating the plants. When the plant or animal dies and decomposes, this nitrogen is put back into the soil in the form of ammonia again (also called **ammonification**), where bacteria will convert it again to nitrites and nitrates. Finally, bacteria convert nitrates into nitrogen gas in a process called **denitrification.**

The water cycle

The final cycle we will look at is the water cycle (Figure 5-12). This cycle is also known as the **hydrologic cycle.**

Figure 5-12 Water cycle.

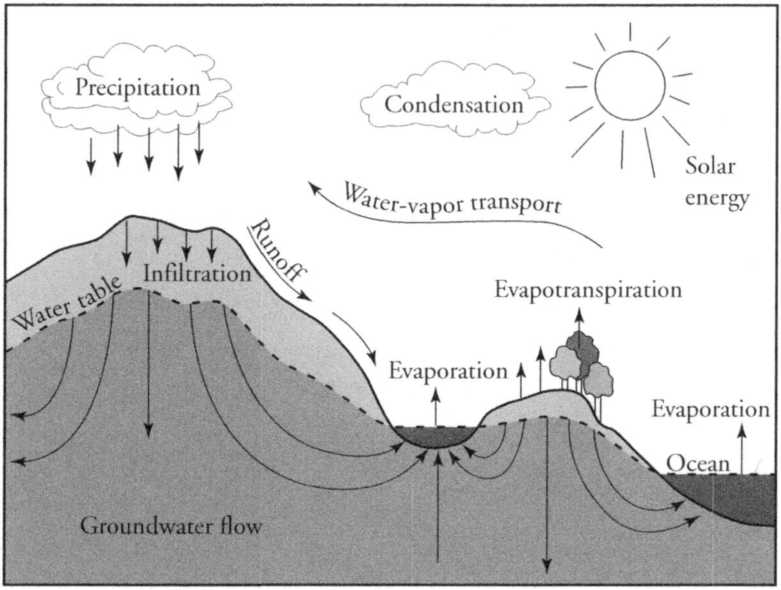

In the water cycle, **precipitation** in the form of rain, snow, sleet, and hail falls to the earth. Some of this precipitation seeps into the earth by **infiltration,** and some makes its way to rivers and lakes in the form of **runoff.** When the water seeps into the earth, it can form in underground pools called **aquifers.**

Evaporation is when the water changes from a liquid to a gas and enters the atmosphere. Water can also enter the atmosphere by evaporating from the cells in plants in a process called **evapotranspiration.** Once in the atmosphere, this water vapor can condense and form clouds in a process known as **condensation.**

In all three cycles, there is no starting or ending point. Matter can be used over and over again in a continuous cycle.

Chapter Check-Out

1. Put the following events in the correct order of primary succession in an ecosystem.

   i. Large plants like shrubs and trees appear.
   ii. A volcano erupts over an area.
   iii. Grasses colonize the area.
   iv. Lichen and moss appear.

   a. i, ii, iii, iv
   b. i, iv, iii, ii
   c. iv, iii, ii, i
   d. ii, iv, iii, i

2. A densely forested area is deforested to make room for human development. Roads are built through the forest, and the area is taken from one large forest to several smaller forests with humans living in between. Which of the following will increase?

   a. the number of predators
   b. the number of prey
   c. the edge habitat
   d. the infiltration rate of water

3. What would happen in the food web below if we removed the red fox?

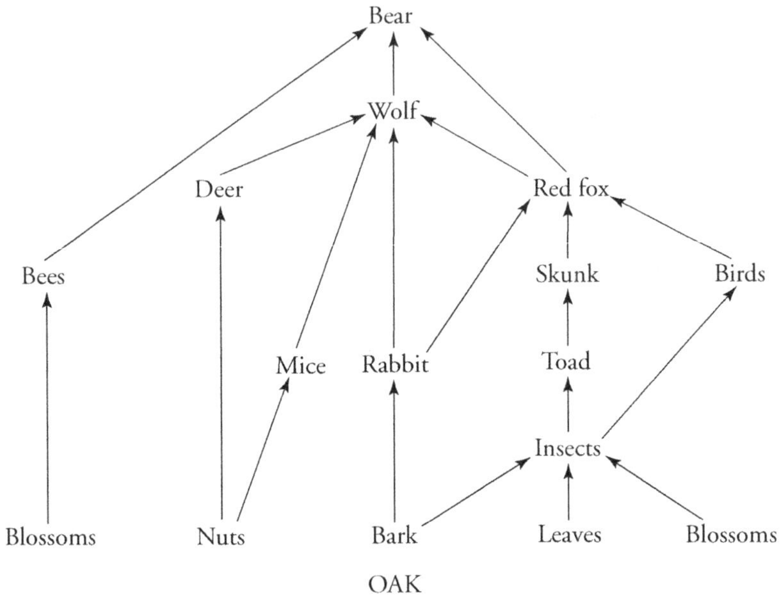

   a. The number of bears would increase.
   b. The number of toads would increase.
   c. The number of rabbits would decrease.
   d. The number of birds would remain the same.

4. What type of organism is missing in the food web shown in question 3?

   a. predators
   b. prey
   c. producers
   d. detritivores

5. In the picture below, you see a sea anemone riding on the shell of a hermit crab. The sea anemone is finding a place to live, and the hermit crab is neither helped nor harmed by the sea anemone. The relationship between the sea anemone and the hermit crab is an example of which of the following?

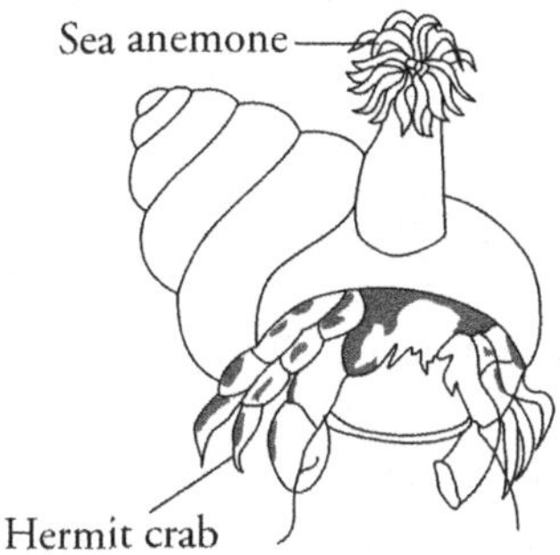

a. commensalism
b. mutualism
c. parasitism
d. predator-prey

6. The nitrogen cycle relies on which of the following organisms in most of the steps of the cycle?

a. primary consumers
b. secondary consumers
c. lichens
d. bacteria

7. Nitrogen and phosphorus are two nutrients that plants need to grow. Farming and ranching practices have allowed too much nitrogen and phosphorus to run off and get into rivers and lakes. When the nitrogen gets into the bodies of water, it can cause an algal bloom. This algal bloom can completely cover the surface of the water, blocking out all sunlight and killing the aquatic organisms below the surface of the water.

   Which of the following is most likely the reason for the death of the aquatic organisms?

   a. Algae are toxic in high numbers.
   b. Loss of light caused a decrease in oxygen below the surface.
   c. Fish that eat the algae became carnivorous.
   d. Nitrogen and phosphorus are absorbed in the skin of aquatic organisms, causing their death.

8. Look at the energy pyramid below. Assume the herbivores contain 1,000 kilocalories. How many kilocalories will be available to the top predator in this diagram?

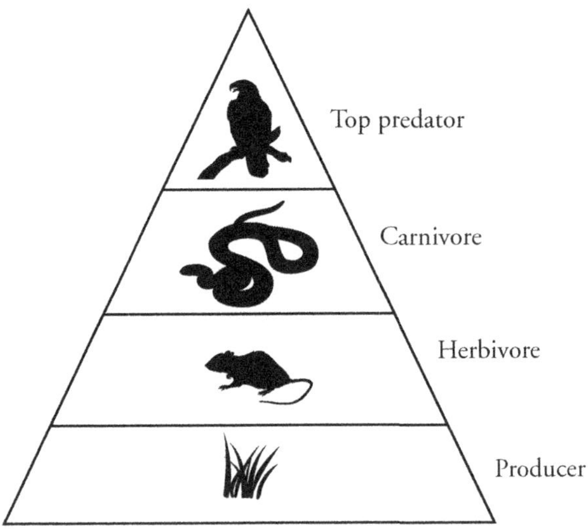

   a. 10,000 kilocalories
   b. 100,000 kilocalories
   c. 10 kilocalories
   d. 100 kilocalories

9. Why do owls have larger eyes than other species of birds?
   a. They are born blind and never completely gain all of their eyesight.
   b. They need large eyes to see predators during the day.
   c. They need large eyes to hunt at night.
   d. They need large eyes to attract members of their species for mating purposes.

10. Which of the following is an example of homeostasis?
   a. A person shivers outside when it is cold.
   b. A person develops a food allergy.
   c. A dog scratches itself due to fleas.
   d. A girl puts her hair in a ponytail when it is hot outside.

Answers: **1.** d **2.** c **3.** b **4.** d **5.** a **6.** d **7.** b **8.** c **9.** c **10.** a

Chapter 6

PRACTICE TEST 1

Directions: Read each question carefully. Choose the best answer from the four answer choices provided.

1. A student finds a picture of the following object in his textbook.

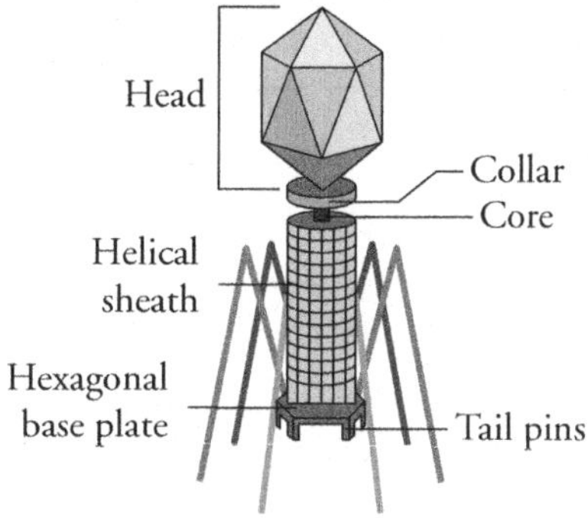

The student would be correct in concluding that the object intends to

a. signal a cell to divide.

b. attach to a cell and inject viral material.

c. move inside a host cell.

d. convert sunlight into energy.

2. There is an active volcano on the island of Hawaii. In 2014, the flow changed course, and instead of the lava flowing into the ocean, as it had historically done, it began to flow onto the land. Put the following descriptions in the correct order of ecological succession that could happen after the lava flow has cooled and hardened.

	Description
I	Moss, lichens, and plants begin to establish the area.
II	Vascular plants begin to appear.
III	There are few nutrients in the lava and sands.
IV	Trees inhabit the island.

   a. I, II, III, IV
   b. II, I, III, IV
   c. III, II, I, IV
   d. III, I, II, IV

3. *Patella vulgata* is a sea snail that has a common name of limpet, or common European limpet. Carolus Linnaeus devised a two-word naming system for all living things. Why does having a universal naming system benefit the scientific community?

   a. Scientists from all over the world can use this system and know exactly which organism they are identifying, regardless of the common name.
   b. Limpets are protected species that the scientific community has deemed ecologically important and worth protecting.
   c. The two-word naming system describes the organism's kingdom and phylum.
   d. The two-word naming system describes the organism's class and order.

4. Electrophoresis is a technique used to separate charged molecules. Electrophoresis separates fragments of DNA by using an electric current to move different-sized molecules through a substance called agar. Since smaller particles move faster than larger molecules, the DNA separates out by size. Electrophoresis permits which of the following?

   a. identifying which DNA is superior
   b. comparing different species and seeing which ones are closely related
   c. extracting carbohydrates from the organism for use in sequencing
   d. classifying organisms based on phenotypic characteristics

5. During the flu season, doctors and nurses at emergency rooms are exposed to the virus many times. What body systems work together to protect the doctors and nurses from these pathogens?

   a. circulatory and endocrine
   b. digestive and nervous
   c. urinary and skeletal
   d. immune and circulatory

6. The primary site of ATP production is the

   a. mitochondria.
   b. cell membrane.
   c. nucleus.
   d. endoplasmic reticulum.

7. The phloem tissue in leaves brings

   a. water from the roots to the rest of the plant.
   b. food from the leaves to the rest of the plant.
   c. phosphorus from the stamen to the anther.
   d. nutrients to the guard cells.

8. It is noticed that a particular gene is increasing in frequency in a population. What would explain this phenomenon?

   a. The population has developed cancer.
   b. Pesticides in the environment have caused a mutation in the population's gene frequency.
   c. The gene must have improved the organism's ability to survive and reproduce.
   d. The number of genes in the population has increased, increasing the population's ability to survive.

9. For many years, ranchers shot and killed wolves to protect their sheep from these predators. After years of this practice, wolves were put on the endangered species list, federally protected, and reintroduced into Yellowstone National Park. Why were the wolves protected?

   a. Wolves were protected to help establish healthy predator-prey relationships in the environment.
   b. It was discovered that the wolves were not the main predators on the farmer's sheep.
   c. The wolves help to offset predation by grizzly bears.
   d. Wolves carry a protozoan in their intestines that, when excreted, has a mutualistic relationship with the fir trees common to the area.

10. In the diagram below, an enzyme is at work on a substrate. Which of the following statements explains what is being shown in the diagram?

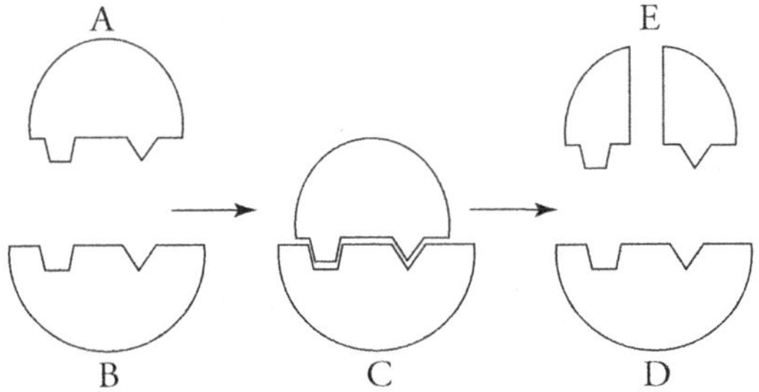

   a. The enzyme is synthesizing two things into one.
   b. The enzyme is reorganizing the cell's genetic matter.
   c. The enzyme is breaking down some chemical bonds of the substrate.
   d. The enzyme is starting the cell cycle for mitosis.

11. Look at the diagram below.

The molecule labeled UACUUCGGCUCAAUU is a strand of

a. DNA.
b. mRNA.
c. tRNA.
d. rRNA.

12. Look at the chemical formula below.

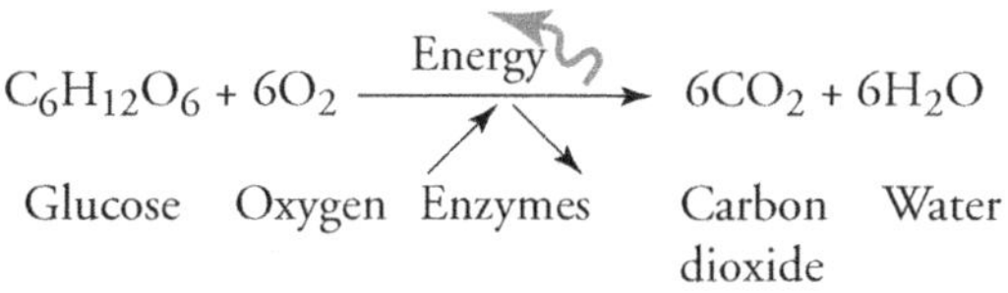

The formula is representing what process?

a. glycolysis
b. Krebs cycle
c. photosynthesis
d. cellular respiration

13. If a green pea plant with genotype GGYy is crossed with a yellow pea plant with genotype Ggyy, what percent of the offspring would have the Ggyy allele combination?

a. 25 percent
b. 50 percent
c. 75 percent
d. 100 percent

14. A food web is diagramed below.

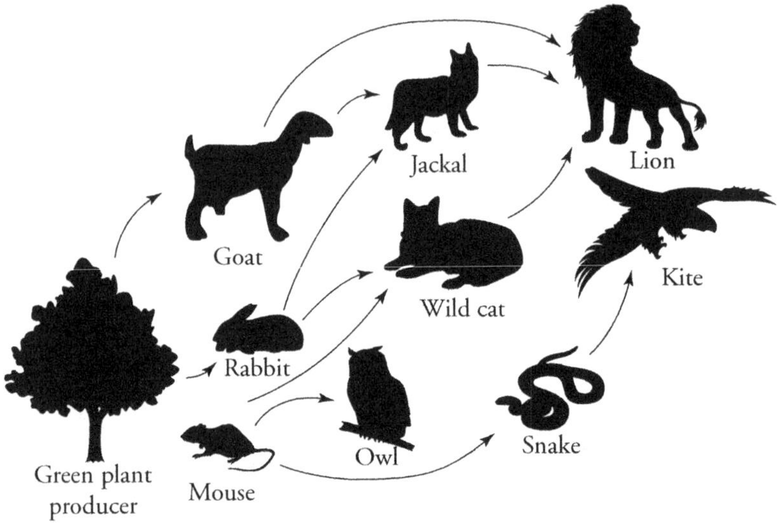

What organisms are missing from the ecosystem?

a. herbivores and producers

b. decomposers and detritivores

c. producers and decomposers

d. secondary and tertiary consumers

15. Why are insects, animals, and the wind important to a plant during pollination?

a. They consume plants, allowing for fungus to repopulate.
b. They trick the plant into thinking that it's time to reproduce.
c. They ensure less diversity of native plant species.
d. They help to increase the genetic variation of the population by aiding in pollen dispersal.

16. Look at the diagram below and identify the stage of mitosis in which the cell is found.

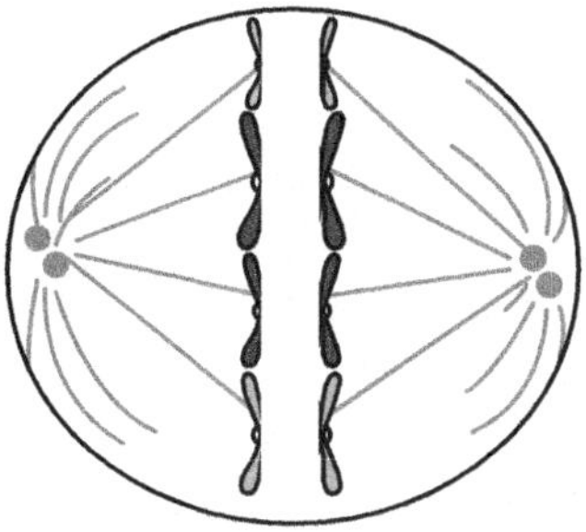

a. prophase
b. metaphase
c. anaphase
d. telophase

17. Diabetes is a disease in which a person has high blood sugar, either because insulin production is weak or because the person's cells do not respond to insulin, or sometimes for both of these reasons. Oftentimes, a person with diabetes will have frequent urination and excessive thirst or hunger. Insulin is a hormone that is normally made in the pancreas and allows a person's body to use sugar from the carbohydrates in food. Insulin is like a key that can "unlock" the cell to allow sugar to enter the cell and be used for energy.

    From the passage above, what would tell a doctor that a patient could have diabetes?

    a. The patient has blood in the urine.
    b. The patient's blood shows a high amount of blood sugar (hyperglycemia).
    c. The patient's cells allow sugar to be used for energy.
    d. The patient complains of vomiting and diarrhea.

18. The bonds that hold together the nucleotides in a strand of DNA are

    a. strong hydrogen bonds.
    b. strong covalent bonds.
    c. weak hydrogen bonds.
    d. weak metallic bonds.

19. A student is trying to identify the following leaf using a dichotomous key.

Dichotomous Key for Leaves

1. Compound or simple leaf
   1a) Compound leaf (leaf divided into leaflets) .... go to step 2
   1b) Simple leaf (leaf not divided into leaflets) .... go to step 4
2. Arrangement of leaflets
   2a) Palmate arrangement of leaflets (leaflets all attached at one central point) ....*Aesculus* (buckeye)
   2b) Pinnate arrangement of leaflets (leaflets attached at several points) .... go to step 3
3. Leaflet shape
   3a) Leaflets taper to pointed tips ....*Carya* (pecan)
   3b) Oval leaflets with rounded tips ....*Robinia* (locust)
4. Arrangement of leaf veins
   4a) Veins branch out from one central point .... go to step 5
   4b) Veins branch off main vein in the middle of the leaf .... go to step 6
5. Overall shape of leaf
   5a) Leaf is heart-shaped ....*Cercis* (redbud)
   5b) Leaf is star-shaped....*Liquidambar* (sweet gum)
6. Appearance of leaf edge
   6a) Leaf has toothed (jagged) edge ....*Betula* (birch)
   6b) Leaf has untoothed (smooth) edge ....*Magnolia* (magnolia)

This leaf is from which plant?

a. *Betula* (birch)

b. *Cercis* (redbud)

c. *Robinia* (locust)

d. *Magnolia* (magnolia)

20. What biomolecules are *most* responsible for storing genetic information?

a. nucleic acids
b. lipids
c. proteins
d. carbohydrates

21. Which amino acid would be translated from mRNA codon CAC?

First letter	Second letter				Third letter
	U	**C**	**A**	**G**	
U	phenylalanine	serine	tyrosine	cysteine	**U**
	phenylalanine	serine	tyrosine	cysteine	**C**
	leucine	serine	stop	stop	**A**
	leucine	serine	stop	tryptophan	**G**
C	leucine	proline	histidine	arginine	**U**
	leucine	proline	histidine	arginine	**C**
	leucine	proline	glutamine	arginine	**A**
	leucine	proline	glutamine	arginine	**G**
A	isoleucine	threonine	asparagine	serine	**U**
	isoleucine	threonine	asparagine	serine	**C**
	isoleucine	threonine	lysine	arginine	**A**
	(start) methionine	threonine	lysine	arginine	**G**
G	valine	alanine	aspartate	glycine	**U**
	valine	alanine	aspartate	glycine	**C**
	valine	alanine	glutamate	glycine	**A**
	valine	alanine	glutamate	glycine	**G**

a. leucine
b. glutamine
c. valine
d. histidine

22. Which two systems work together when a person goes on a jog and begins to sweat?

a. digestive and circulatory
b. integumentary and excretory
c. skeletal and nervous
d. urinary and endocrine

23. A student looks under the microscope and sees an organism that has membrane-bound organelles, a nuclear membrane and nucleus, and flagella. The student concludes that the organism is

a. eukaryotic.
b. prokaryotic.
c. a fungus.
d. a plant.

24. There is a protozoan that is found in the abdomen of termites that helps them to digest wood. Termites live on cellulose from the wood they eat, and depend on the protozoa in their guts to provide the enzymes that can digest the wood. The relationship between the termite and the protozoan is an example of which of the following?

a. mutualism
b. commensalism
c. parasitism
d. predator/prey

25. Two organisms are found to have similarities in their DNA. What would this tell you about these two organisms?

a. The organisms are plants.
b. The organisms share a common ancestry.
c. The organisms have cellular mutations.
d. The organisms are prokaryotic.

26. Meiosis contributes to genetic variation in populations. Which of the following best describes this occurrence?

 a. Meiosis is part of the cell cycle, increasing genetic variation.
 b. Meiosis produces the reproductive cells that yield individuals that are different from both their parents.
 c. Meiosis produces diploid cells.
 d. Meiosis guarantees the benefits of asexual reproduction.

27. How does the burning of fossil fuels contribute to a net increase of carbon in the atmosphere?

 a. The burning of fossil fuels releases sulfur dioxide into the atmosphere; sulfur dioxide is a major component of acid rain.
 b. The burning of fossil fuels leads to eutrophication, which is a major problem in the world's rivers and lakes.
 c. Carbon was stored underground for millions of years in the form of coal, oil, and natural gas. With the burning of fossil fuels, this carbon is now added to the carbon cycle in the atmosphere.
 d. Carbon releases ozone, which causes lung irritation due to the destruction of the stratospheric ozone layer.

28. According to the table below, which of these organisms are most closely related to each other?

	Organism 1	*Organism 2*	*Organism 3*	*Organism 4*
Kingdom	Plantae	Plantae	Plantae	Plantae
Phylum	Pinophyta	Magnoliophyta	Magnoliophyta	Magnoliophyta
Class	Pinopsida	Lilliopsida	Lilliopsida	Magnoliopsida
Order	Pinales	Zingiberales	Commelinales	Fageles
Family	Pinaceae	Musaceae	Poaceae	Fagaceae

 a. 1 and 2
 b. 2 and 4
 c. 2 and 3
 d. 3 and 4

29. In a cell, protein synthesis is the primary function of what organelle?

a. ribosomes
b. mitochondria
c. chlorophyll
d. nucleus

30. Below is a diagram of a building block of DNA. What is the name of this part of DNA?

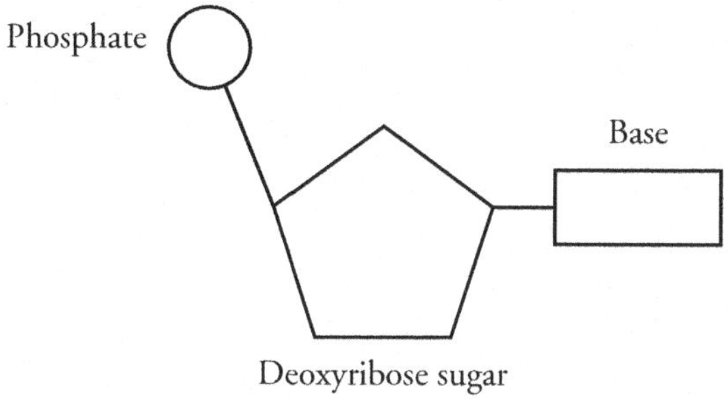

a. nucleolus
b. mitochondria
c. carbohydrate
d. nucleotide

31. A farmer is struggling to control an insect pest on his farm. He researches the insect and discovers that a particular bird is a predator of this insect. The farmer purchases a number of these birds and releases them into the area. What is a possible negative consequence of the farmer's actions?

a. The birds consume beneficial insects, as well as the pest species.
b. The birds will increase biodiversity of native plant species.
c. The birds will increase fragmentation and decrease edge effect.
d. The birds pollinate native species, increasing genetic variation.

32. A plant is found lying on its side as shown in the diagram below.

The plant is still found growing upward because of

a. phototropism.
b. gravitropism.
c. thigmotropism.
d. transpiration.

33. What is the location of the chromosomes during metaphase?

a. at the poles
b. attached to centromeres and being pulled toward opposite sides of the cell
c. lined up along the equator of the cell
d. in two completely separate cells

34. The process in which organisms that are best adapted to the environment survive and reproduce is known as

a. punctuated equilibrium.
b. artificial insemination.
c. gradualism.
d. natural selection.

35. Sunlight, climate, rainfall, minerals, and gases are all abiotic factors that relate to a(n)

a. ecosystem.
b. biosphere.
c. organism.
d. organelle.

36. Which of the following represents a carbohydrate molecule?

a.

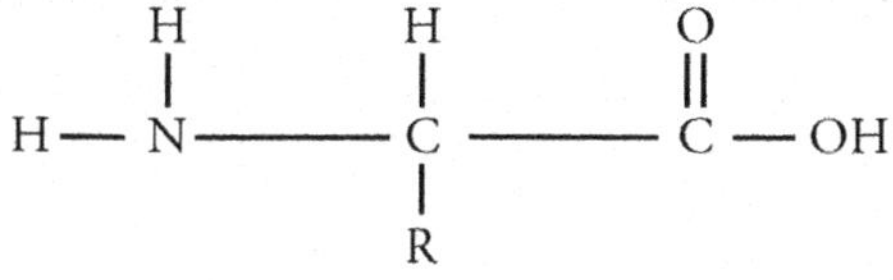

b.

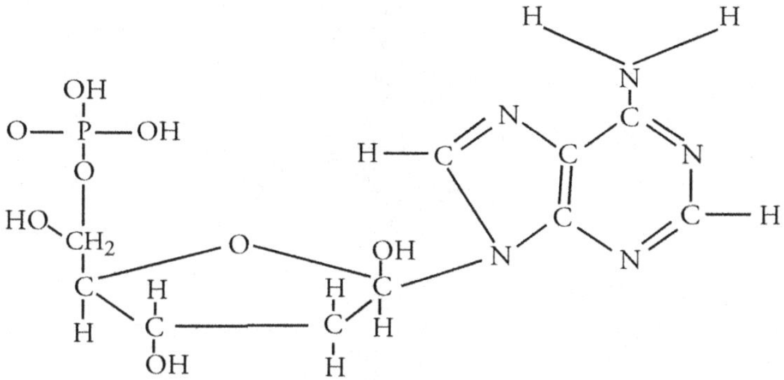

c.

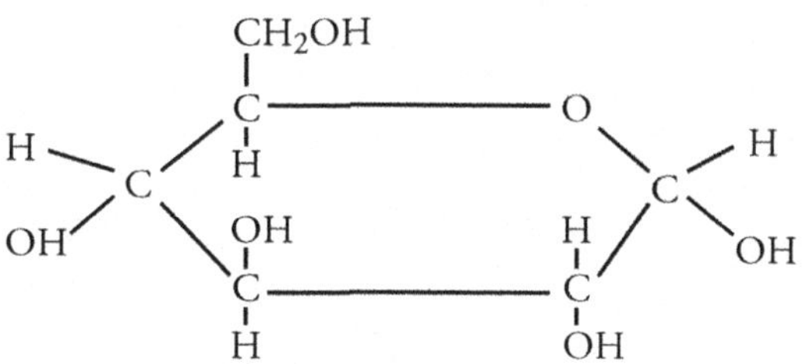

d.

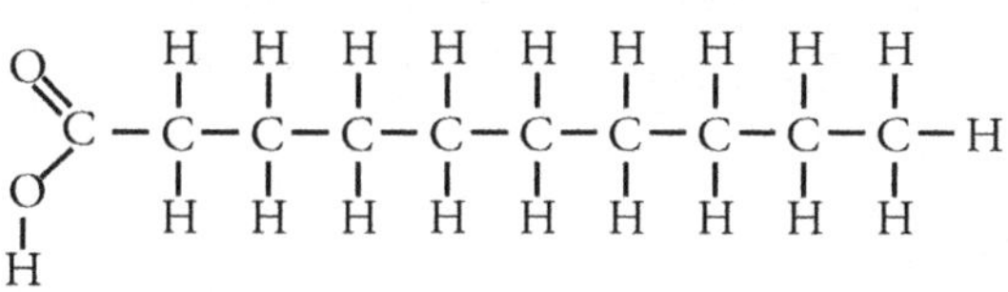

37. Look at the energy pyramid below. If the top predator in the diagram has 10 kilocalories of energy, how many kilocalories of energy did the producers contain?

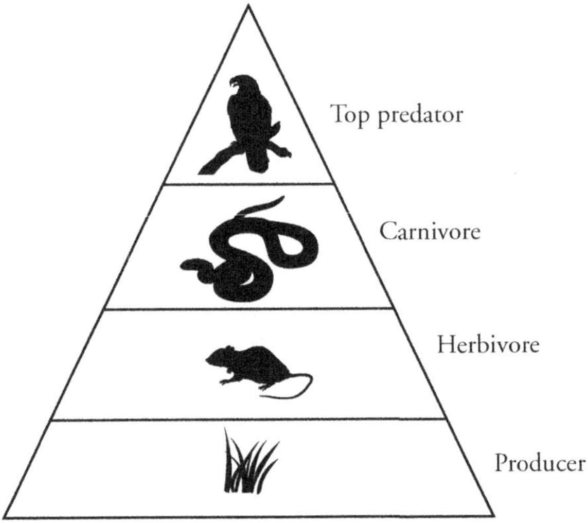

a. 1,000 kilocalories

b. 10,000 kilocalories

c. 40 kilocalories

d. 1 kilocalorie

38. In humans, red hair (b) is recessive over brown hair (B). A person who is heterozygous for brown hair (Bb) marries a homozygous redhead (bb). Which of the following genotypic ratios can be expected?

a. 0 BB : 2 Bb : 2 bb

b. 2 BB : 2 Bb : 0 bb

c. 2 BB : 0 Bb : 2 bb

d. 1BB : 2 Bb : 1 bb

39. The kit fox and the red fox have anatomical structures that are similar enough to give scientists evidence that they came from a common ancestor. However, the kit fox is sandy-colored and the red fox has coloring to blend in with a forested area. The adaptations that these two foxes have developed in their coloring have allowed them to

a. better find a mate.

b. attract prey.

c. consume a variety of food items.

d. blend better into their habitats.

40. Decomposers are important organisms in the environment. One of the main roles of decomposers is to

a. be a food item for herbivores.

b. allow carnivores to find their specific niches.

c. break down dead or dying organisms.

d. allow glucose to be converted to nitrogen.

41. Uncontrollable cell growth is the definition of cancer. Below is a diagram of the cell cycle.

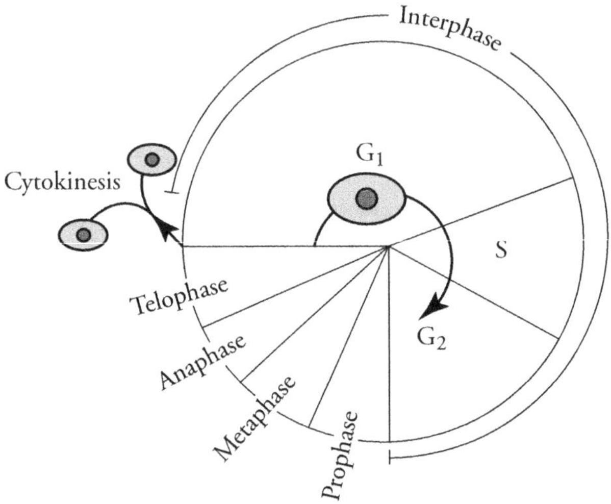

Which of the following descriptions correctly describes what happens to a cell when it becomes cancerous?

a. The cell stays for too long in the G_1 phase.
b. The cell does not complete the G_2 phase.
c. The cell cycle continues without regulation.
d. The cell remains in the S phase twice as long as the other phases.

42. A student decides to build a pond in her backyard under a large oak tree. She digs a 3-foot-deep hole to form the pond, lines the pond with a plastic liner, fills the pond with water, and adds plants and fish. For the first few years, the pond thrives, and the student enjoys it. However, 5 years down the road, she notices that her 3-foot-deep pond is now only 1 foot deep, and the fish are beginning to struggle for room. She investigates and discovers that 5 years of the oak tree dropping leaves in the pond is causing the pond to fill up from the bottom with mud and muck. What disturbance is occurring in the pond?

a. succession
b. eutrophication
c. mutualism
d. deforestation

43. Which of the following molecules is responsible for taking a copy of the instructions coded for by a gene from the nucleus into the cytoplasm and to the ribosomes, the protein factories of our cells?

a. DNA

b. mRNA

c. rRNA

d. tRNA

44. A diagram of the reproductive parts of a flower is shown below. What is the purpose of the stamen?

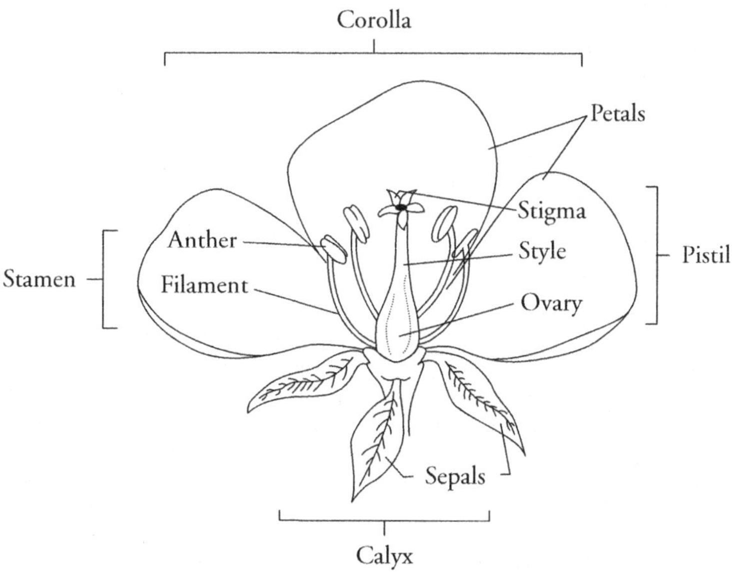

a. The stamen is the female reproductive structure.

b. The stamen attracts pollinators to the flower.

c. The stamen supports the sepals.

d. The stamen is the male reproductive structure.

45. There are two different ways that viruses can reproduce. In the lytic cycle, the virus injects viral DNA into the host and allows the host to make copies of the virus. In the other cycle, the virus injects its DNA into the host cell, causing the viral DNA to combine with the host's DNA. The virus can later remove from the host DNA and cause the host to begin making copies and producing new viruses. What is the name of this second type of viral reproduction?

   a. the lytic cycle
   b. the viral cycle
   c. the lysogenic cycle
   d. the reproductive cycle

46. Look at the pictures below of different plant species. Which plant would you expect to be growing on the floor of the rain forest where there is little light due to large trees in the canopy?

   a. 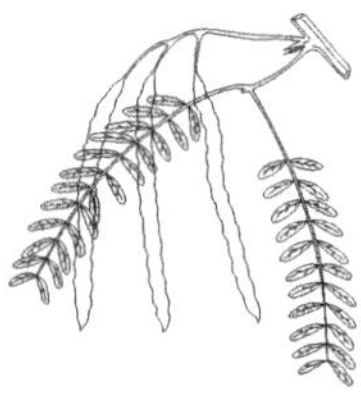

   b. 

   c. 

   d. 

47. Your body has some cells used for eyesight, other cells to make your heart beat, and other cells that are found in your skin for protection. The differentiation of cells in your body is directed by your

   a. DNA.
   b. RNA.
   c. mitochondria.
   d. pituitary gland.

48. Your skin, hair, and nails protect your body from outside pathogens and are part of what body system?

a. digestive
b. endocrine
c. integumentary
d. cardiovascular

49. The link between the nucleotide sequence of a gene and the amino acid sequence of the protein specified by the gene is the

a. protein.
b. genetic code.
c. carbohydrate-gene link.
d. chromosome.

50. Zebra mussels are an invasive species that were introduced into the Great Lakes from the ballast water of ships. This species has been able to spread from the Great Lakes into many different water sources because, like many invasive species, it

a. can outcompete native species.
b. is found naturally in the lakes and rivers of North America.
c. can live in and out of a water source.
d. can go into hibernation and wait until conditions are more favorable.

51. During a drought, a plant is able to close its guard cells to help prevent transpiration. This ability of the plant helps to

a. remove excess carbon dioxide from its cells.
b. keep the stomata from dividing.
c. increase the amount of oxygen the plant needs for reproduction.
d. maintain homeostasis.

52. The study of homologous structures in mature organisms provides evidence for

a. DNA cloning.
b. understanding predator-prey relationships.
c. a common ancestor.
d. how enzymes can be found in all life forms.

53. Mutations can lead to evolutionary change because they can

a. lead to mass extinctions.

b. contribute to new variations in organisms.

c. cause cancer cells to develop.

d. be transferred to other species in the environment.

54. A cladogram like the one below is used to

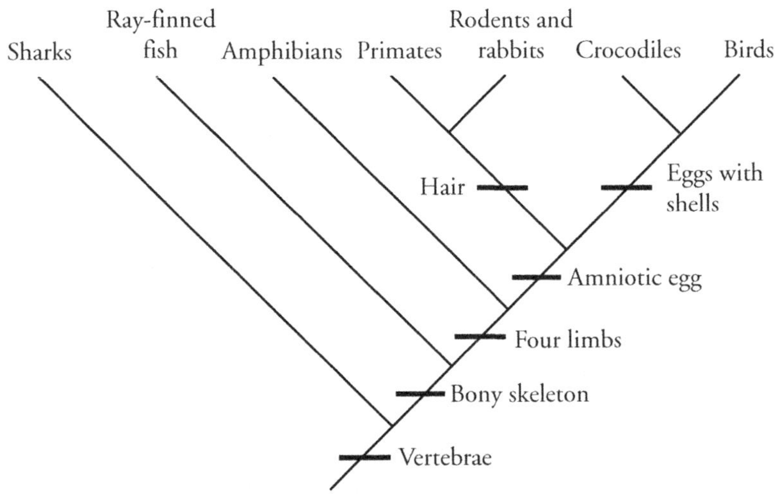

a. convey information about ancestors and descendants.

b. allow early taxonomists to understand organisms' DNA.

c. identify the difference between plants and animals.

d. identify mammals from other animal species.

Answers

1. b
2. d
3. a
4. b
5. d
6. a
7. b
8. c
9. a
10. c
11. b
12. d
13. a
14. b
15. d
16. c
17. b
18. b
19. a
20. a
21. d
22. b
23. a
24. a
25. b
26. b
27. c
28. c
29. a
30. d
31. a
32. b
33. c
34. d
35. a
36. c
37. b
38. a
39. d
40. c
41. c
42. a
43. b
44. d
45. c
46. d
47. a
48. c
49. b
50. a
51. d
52. c
53. b
54. a

Chapter 7

PRACTICE TEST 2

Directions: Read each question carefully. Choose the best answer from the four answer choices provided.

1. The protein cytochrome c, essential for aerobic respiration, is found in all aerobic organisms. The fact that this protein is found universally is evidence that

   a. all organisms are aerobic.

   b. all aerobic organisms descended from a common ancestor.

   c. anaerobic organisms must have descended from aerobic organisms.

   d. cytochrome c is an enzyme found in living organisms.

2. Below is a diagram of specialized cells, known as guard cells, on the surface of a leaf. What is the main function of these cells?

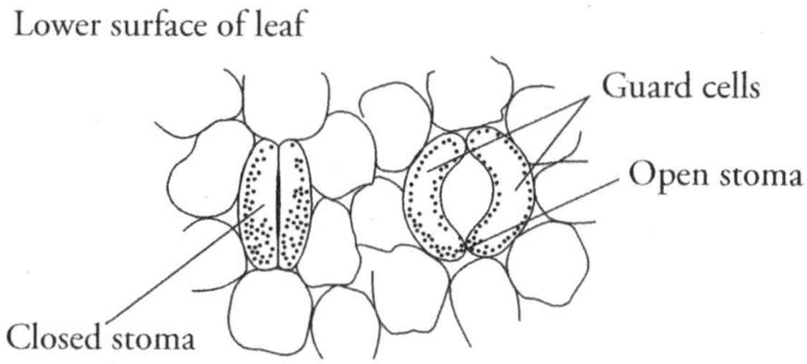

   a. Guard cells are involved in sexual reproduction.

   b. Guard cells perform photosynthesis.

   c. Guard cells are specialized xylem tissue.

   d. Guard cells change the size of the stomata to lessen water loss.

3. Slender-horned gazelles are grazing animals that are native to the Sahara Desert. Which animal can be predicted to have the most similar adaptations to those of a slender-horned gazelle?

   a. a grazing animal native to the Chihuahuan Desert
   b. a predator of a slender-horned gazelle
   c. an animal with a mutualistic relationship to gazelles
   d. an animal found only in taiga ecosystems

4. If during DNA replication a base pair is changed as shown in the diagram below, what could be the result?

G G A	C [T] C	C T C
C C T	G [A] G	G A G

Before

G G A	C [G] C
C C T	G [C] G

After

   a. a new species
   b. an identical twin
   c. a mutation
   d. fertilization

5. The cladogram shows the evolution of different animal species from the fossil record.

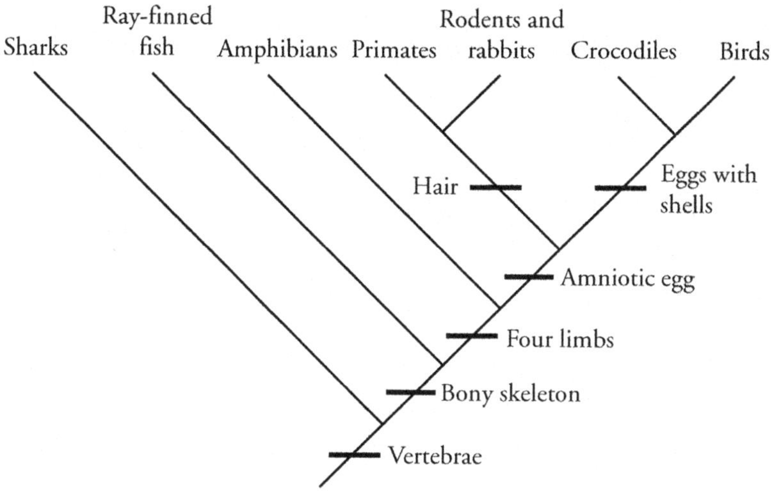

Which of the following scenarios would change the legitimacy of this cladogram?

a. a ray-finned fish with four limbs
b. an amphibian with vertebrae
c. a shark with vertebrae
d. a rodent with an amniotic egg

6. Uncontrolled cell growth in a person's body can restrict the normal functioning of the surrounding cells. This situation is known as

a. transcription.
b. cancer.
c. genetic drift.
d. translation.

7. Transportation of water from the roots to the leaves of a plant would be interrupted if which of the following happened?
    a. The stomata were closed.
    b. The phloem was damaged.
    c. Nitrogen was absent in the soil.
    d. The xylem was damaged.
8. Antibodies are made of amino acids, which means they would belong to the class of biomolecules known as
    a. proteins.
    b. nucleic acids.
    c. carbohydrates.
    d. lipids.
9. What could contribute to gene flow between populations?
    a. genetic drift
    b. migration
    c. mutation
    d. natural selection

10. Many people tend to over-fertilize their lawns in the spring, hoping for a beautiful lawn. When it rains, the fertilizers can wash off the lawns and into the neighboring rivers and lakes and can lead to rapid growth of algae, a process known as eutrophication. What does this increased growth of algae do to the freshwater ecosystem?

a. The algae are toxic and cause extensive fish poisoning.

b. The algae add zinc to the water, leading to an abundance of microorganisms.

c. The algae deplete the nutrients in the water and reduce the resources that are available to other aquatic organisms.

d. The algae mutate and begin consuming freshwater microorganisms.

11. You find a butterfly that is a solid gray color with no dark or white on the wings. Using the chart below, what family would you put it in?

Organism	1	2	3	4
Common name	Orange-barred sulphur	Orange-banded protea	Silver-spotted flambeau	Silver-studded blue
Class	Insecta	Insecta	Insecta	Insecta
Order	Lepidoptera	Lepidoptera	Lepidoptera	Lepidoptera
Family	Peridae	Lycaenidae	Nymphalidae	Lycaenidae
Genus	*Phoebis*	*Capys*	*Agraulis*	*Plebejus*

a. Insecta

b. Lycaenidae

c. Peridae

d. Nymphalidae

12. What concept is best illustrated by the picture below?

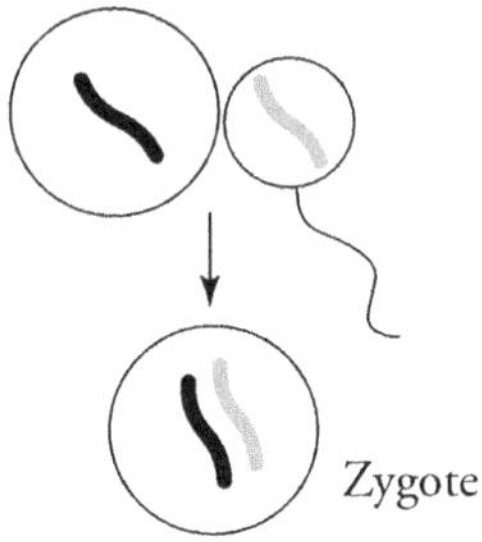

a. Asexual reproduction
b. During the telophase stage of mitosis, the individual zygotes become one.
c. The cell cycle continues without regulation.
d. During sexual reproduction, the offspring receive the same number of chromosomes from each parent.

13. You eat dinner after school one day. What two body systems are working together to get the nutrients from your food to your brain?

a. respiratory and digestive
b. immune and endocrine
c. circulatory and endocrine
d. circulatory and digestive

14. The nitrogen cycle is a complex way that nitrogen is able to move around in the environment. What role do the nitrogen-fixing bacteria serve in the nitrogen cycle?

a. They convert phosphorus into nitrogen.
b. They convert nitrogen into a form that is usable for plants.
c. They transport water from roots to stems to leaves in plants.
d. They fix oxygen for use in photosynthesis.

15. Viruses are different from living organisms in that viruses

a. cannot reproduce on their own.
b. are larger than cells.
c. need oxygen to reproduce.
d. require water for homeostasis.

16. As organisms move up the energy pyramid, fewer kilocalories are available. This is because

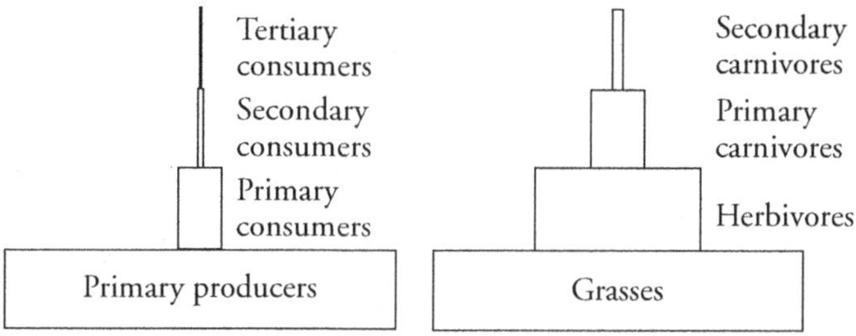

Pyramid of Energy **Pyramid of Biomass**

a. there are more organisms at the top of the food chain.
b. some energy is converted to heat.
c. larger animals need fewer kilocalories.
d. tertiary consumers are herbivores.

17. Healthcare workers who take X-rays for a living must protect themselves from the radiation that is emitted. Radiation can penetrate cells and cause damage to the cells' DNA.

Changes to DNA are an example of which of the following?

a. erythroblast differentiation
b. sexual reproduction
c. protein synthesis
d. mutation

18. One important function that bones do for the body is to

   a. provide a waterproof layer for protection.
   b. allow gas exchange.
   c. produce blood cells.
   d. produce toxins that weaken immune cells.

19. What structure is common to all types of cells?

   a. cell membrane
   b. chloroplast
   c. cell wall
   d. flagella

20. Charles Darwin studied many different types of finches on his trip to the Galapagos Islands. He discovered that on different islands there lived different finches, all of which had adapted to different environmental niches. These finches were unable to interbreed. Darwin concluded that these finches were

   a. not related.
   b. different species.
   c. homozygous recessive.
   d. apex predators.

21. A person goes on a run on a hot summer day. After just a few minutes she is sweating and breathing harder, and her heart rate has increased. Her body's response to her increased physical activity is to

   a. collapse.
   b. stop non-essential body processes.
   c. keep all her blood in her abdomen.
   d. maintain homeostasis.

22. During the final stage of cell division, the chromosomes assemble at the two poles and begin to uncoil, a nuclear envelope re-forms around each chromosome, and cytokinesis follows. What is this phase of cell division?

a. prophase
b. metaphase
c. anaphase
d. telophase

23. A clown fish is able to live in the tentacles of a sea anemone and not be harmed by the anemone's stinging tentacles. The clown fish gets protection from the anemone, and the anemone obtains nutrients from the clown fish's waste. What kind of symbiotic relationship does this describe?

a. mutualism
b. commensalism
c. parasitism
d. predator-prey

24. Tall pea plants have the allele (T) and short pea plants have the allele (t). Round seeds have the allele (R), which is dominant to wrinkled seed plants with the allele (r). What cross would give a short pea plant with a wrinkled seed?

a. TtRr × Ttrr
b. TTRR × TtRr
c. TtRR × ttRR
d. TTRr × TTRr

25. After a volcanic eruption, an area is covered with molten lava. This lava cools, hardens, and becomes rock. Lichen will begin to cover the rock, soil will form, and eventually small plants will occupy the area, followed by larger plants. What is the name of this process?

    a. commensalism
    b. succession
    c. eutrophication
    d. anthropogenic

26. Which two systems of the human body are primarily used to fight pathogens?

    a. endocrine and integumentary
    b. lymphatic and nervous
    c. immune and lymphatic
    d. skeletal and endocrine

27. If a large part of a cell's ribosomes are found attached to the endoplasmic reticulum, the cell is specialized to

    a. make mitochondria.
    b. build cell walls.
    c. perform photosynthesis.
    d. manufacture proteins.

28. A person has skin cells and heart cells that contain the same genetic sequences. However, the cells are different because the skin cells

a. are haploid and the heart cells are diploid.
b. use different genes than the heart cells.
c. have cell walls and the heart cells do not.
d. do not contain nuclei.

29. Look at the diagram below and identify the process that is being shown.

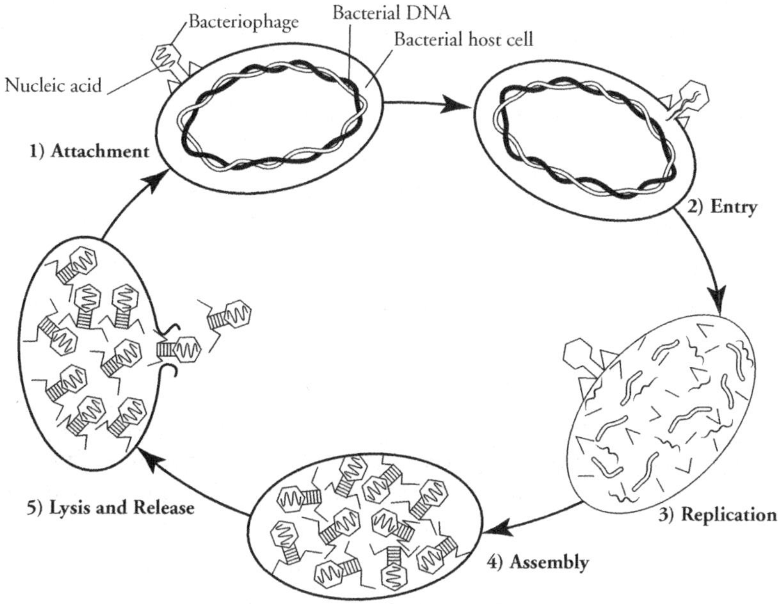

a. lytic cycle
b. lysogenic cycle
c. cell cycle
d. Krebs cycle

30. The role of an enzyme in a chemical reaction is to

a. prevent the reaction.
b. slow down the reaction.
c. speed up the reaction.
d. change the direction of the reaction.

31. A karyotype can be used to diagnose which of the following conditions?

a. a viral infection
b. a bacterial infection
c. any non-genetic disorder
d. an incorrect number of chromosomes

32. You go to the kitchen and accidently touch a hot pan on the stove. You immediately jerk your hand back in pain. What two body systems were working together to protect you from getting burned?

a. circulatory and immune
b. integumentary and skeletal
c. muscular and endocrine
d. nervous and muscular

33. A cell that is 90 percent water and 10 percent solute is placed in a hypertonic solution that is 80 percent water and 20 percent solute. If the cell membrane is impermeable to the solute, what will happen to the cell after a few hours?

a. Water will move out of the cell and into the solution.
b. Water will move into and out of the cell equally.
c. Water will move into the cell and out of the solution.
d. Solute will move out of the cell and into the solution.

34. In Texas, many oak trees are covered with medium-sized plants known as ball moss. These plants grow up in the tops of the trees to gain light. In most situations, the tree is not bothered by the ball moss because the moss does not take away any of the tree's nutrients, water, or light. What type of symbiotic relationship is described by the oak trees and the ball moss?

a. mutualism

b. commensalism

c. parasitism

d. predator-prey

35. Why might identical twins, whose genes are exactly the same, look different from each other?

a. The loci on the genes are inverted.

b. The twins exhibit sexual dimorphism.

c. Some gene expression may be changed by environmental factors.

d. There is a conditioned response to prevent polydactyly.

36. Look at the food web pictured below.

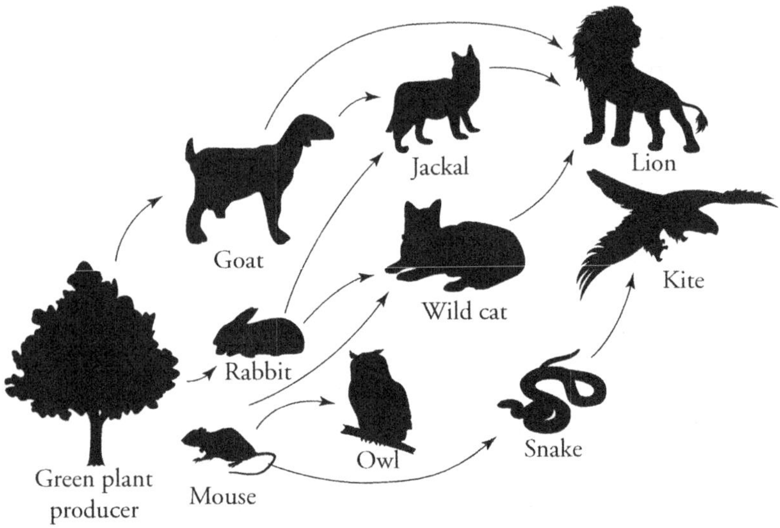

What organism is acting as a tertiary consumer?

a. kite
b. mouse
c. owl
d. snake

37. When it is hot and dry outside, the pores on the surfaces of a leaf will close to prevent water loss during the day. What is the name of the pores that open and close depending on the external conditions and the amount of water a plant needs?

a. vacuole
b. stoma
c. Golgi apparatus
d. nucleolus

38. The scientific name for a common house cat is *Felis catus.* Which of the following organisms would be the most closely related to the common house cat?

a. *Canus catus*
b. *Felis silvestris*
c. *Toxunda catus*
d. *R. catusulus*

39. When a mistake occurs during DNA replication and your body's safeguards do not stop it, there can be a permanent change to your DNA. What is the name of this phenomenon?

a. transcription error
b. genetic modification
c. meiosis II
d. genetic mutation

40. The finches that Charles Darwin studied on the Galapagos Islands had many different beaks. Some of the beaks were long and narrow, while others were short and powerful. Darwin concluded that the different sizes and shapes of the beaks were to

a. provide reproductive advantage.
b. differentiate species.
c. limit competition for scarce nutrients.
d. help in displays of dominance.

41. As we burn fossil fuels, we are taking carbon from deep below the ground and adding it to the atmosphere. The carbon that makes up our fossil fuels was carbon that was actively cycling in the atmosphere millions of years ago but was buried and stored until the present. How does this process affect the carbon cycle today?

a. There is a net increase in atmospheric carbon that can lead to climate change.
b. It allows for atmospheric carbon to undergo sedimentation.
c. It reduces the nutrients in the soil.
d. It releases heat from photosynthetic organisms.

42. A beetle population that was once a single species is divided and isolated from one another by a river. This river effectively splits the population into two. After many generations, each population evolves genetic differences so that they can no longer interbreed. What is this referring to?

   a. the bottleneck effect
   b. behavioral isolation
   c. temporal isolation
   d. geographic isolation

43. If a heterozygous purple-flowered pea plant (Bb) is crossed with another heterozygous purple-flowered pea plant (Bb), what is the expected genotypic outcome of the crossing?

   a. 50 percent (BB); 25 percent (Bb), 25 percent (bb)
   b. 25 percent (BB); 25 percent (Bb), 50 percent (bb)
   c. 25 percent (BB); 50 percent (Bb), 25 percent (bb)
   d. 50 percent (BB); 50 percent (Bb), 0 percent (bb)

44. What biomolecules are generally found in a 1:2:1 ratio and are an important energy source for living things?

   a. carbohydrates
   b. lipids
   c. nucleic acids
   d. proteins

45. The pine bark beetle is an insect that is destroying millions of acres of forest in western North America. The outbreak is thought to be a consequence of global warming. Which of the following could be a consequence of an infestation of pine bark beetles?

   a. an increase in precipitation
   b. a decrease in forest fires
   c. an increase in mudslides
   d. a decrease in pine bark beetle predators

46. Animals must release the chemical energy from sugars to make the energy they need to grow, reproduce, and survive. This complex process is known as

a. photosynthesis.
b. cellular respiration.
c. transpiration.
d. binary fission.

47. A student looks under a microscope and observes a cell with chromosomes lined up down the middle as shown in the picture below.

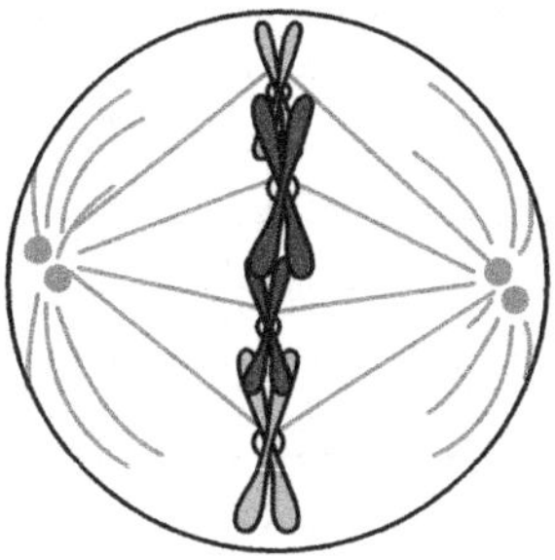

What stage of mitosis is shown?

a. prophase
b. metaphase
c. anaphase
d. telophase

48. Use the codon chart below to name the amino acid that codes for UGC.

First letter	Second letter				Third letter
	U	C	A	G	
U	phenylalanine	serine	tyrosine	cysteine	U
	phenylalanine	serine	tyrosine	cysteine	C
	leucine	serine	stop	stop	A
	leucine	serine	stop	tryptophan	G
C	leucine	proline	histidine	arginine	U
	leucine	proline	histidine	arginine	C
	leucine	proline	glutamine	arginine	A
	leucine	proline	glutamine	arginine	G
A	isoleucine	threonine	asparagine	serine	U
	isoleucine	threonine	asparagine	serine	C
	isoleucine	threonine	lysine	arginine	A
	(start) methionine	threonine	lysine	arginine	G
G	valine	alanine	aspartate	glycine	U
	valine	alanine	aspartate	glycine	C
	valine	alanine	glutamate	glycine	A
	valine	alanine	glutamate	glycine	G

a. phenylalanine
b. glutamine
c. tryptophan
d. cysteine

49. Each enzyme will speed up one particular reaction. This is because enzymes are

a. specific to the substrate they bind with.
b. limited resources.
c. infectious in nature.
d. found in low-pH environments.

50. The theory of natural selection states that

a. organisms can chose to evolve.
b. asexual reproduction is necessary for speciation.
c. organisms better adapted to their environment have greater reproductive success.
d. organisms are bred for desirable traits.

51. Plants have chloroplasts that help them produce their own energy by converting sunlight to sugar. What is the name of this process?

a. photosynthesis
b. cellular respiration
c. turgor pressure
d. translocation

52. In a DNA double helix, the two DNA chains are held together by

a. strong hydrogen bonds.
b. weak hydrogen bonds.
c. strong ionic bonds.
d. weak ionic bonds.

53. Ecological succession is the

a. alteration of an organism over time.
b. diversity within a community.
c. process of toxins building up in the food chain.
d. gradual progression over time from one biotic community to another.

54. Use the dichotomous key to identify Bird X.

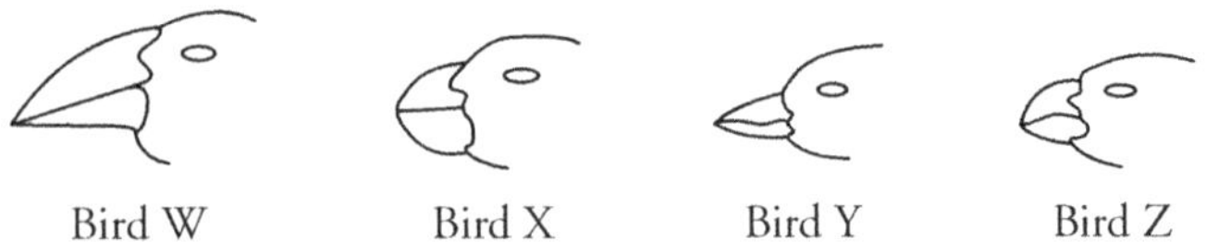

Dichotomous Key to Representative Birds
1. a. The beak is relatively long and slender .... *Certhidea*
b. The beak is relatively stout and heavy .... go to 2
2. a. The bottom surface of the lower beak is flat and straight .... *Geospiza*
b. The bottom surface of the lower beak is curved .... go to 3
3. a. The lower edge of the upper beak has a distinct bend .... *Camarhynchus*
b. The lower edge of the upper beak is mostly flat .... *Platyspiza*

a. *Certhidea*
b. *Geospiza*
c. *Camarhynchus*
d. *Platyspiza*

Answers

1. b
2. d
3. a
4. c
5. a
6. b
7. d
8. a
9. b
10. c
11. c
12. d
13. d
14. b
15. a
16. b
17. d
18. c
19. a
20. b
21. d
22. d
23. a
24. a
25. b
26. c
27. d
28. b
29. a
30. c
31. d
32. d
33. a
34. b
35. c
36. a
37. b
38. b
39. d
40. c
41. a
42. d
43. c
44. a
45. c
46. b
47. b
48. d
49. a
50. c
51. a
52. b
53. d
54. d

Made in the USA
Coppell, TX
02 April 2023

15105813R00108